FIRST LIFE

THE PUBLISHER GRATEFULLY ACKNOWLEDGES THE GENEROUS SUPPORT OF THE GENERAL ENDOWMENT FUND OF THE UNIVERSITY OF CALIFORNIA PRESS FOUNDATION.

FIRST LIFE

Discovering the Connections between Stars, Cells, and How Life Began

David Deamer

UNIVERSITY OF CALIFORNIA PRESS

Berkeley   Los Angeles   London

University of California Press, one of the most distinguished university presses in the United States, enriches lives around the world by advancing scholarship in the humanities, social sciences, and natural sciences. Its activities are supported by the UC Press Foundation and by philanthropic contributions from individuals and institutions. For more information, visit www.ucpress.edu.

University of California Press
Berkeley and Los Angeles, California

University of California Press, Ltd.
London, England

Library of Congress Cataloging-in-Publication Data

Deamer, D. W.
  First life : Discovering the Connections between stars, cells, and how life began / David Deamer.
    p. cm.
  Includes bibliographical references and index.
  ISBN 978-0-520-25832-7 (cloth : alk. paper)
  1. Exobiology. 2. Life—Origin. 3. Evolution (Biology) I. Title.
  QH326.D43 2011
  576.8'3—dc22

                                                2010035393

Manufactured in the United States of America
19 18 17 16 15 14 13 12 11
10 9 8 7 6 5 4 3 2 1

The paper used in this publication meets the minimum requirements of ANSI/NISO Z39.48-1992 (R 1997) (*Permanence of Paper*).

Cover illustration: (Upper) Membrane structures produced by self-assembly of organic compounds extracted from the Murchison carbonaceous meteorite, photo by the author. (Lower) Myelin figures growing out of a dried phospholipid sample interacting with water, photo courtesy of Alison North. The membranous structures were visualized by light microscopy and false color was added to improve contrast.

CONTENTS

ACKNOWLEDGMENTS

How life began is one of the outstanding questions of science, yet it is surprising how few researchers give it serious attention. This becomes abundantly clear if we consider scientific meetings dedicated to other disciplines. For instance, the annual meetings of chemists, biophysicists, and neurobiologists attract 10,000 or more researchers to cities like Chicago, Boston, and San Francisco, but the international meeting of origins of life researchers is lucky to have 500 participants show up in places like Florence, Oaxaca, and Montpelier. This means that virtually everything I discuss in this book comes from the work of just a hundred or so fellow scientists. Here I want to thank those friends and colleagues who have shared their knowledge with me over the past 30 years, who in a sense are coauthors of this book.

David Cornwell at The Ohio State University College of Medicine guided my doctoral research and introduced me to the wonders of lipid chemistry. I will not soon forget the pleasure of extracting pure phosphatidylcholine from the yolks of a dozen eggs, then seeing how it behaved on a homemade Langmuir trough.

Dan Branton and Lester Packer hosted my postdoctoral research at UC Berkeley and introduced me to a level of science that is unique to the great research universities of the world. With them, I learned about the structure and function of biological membranes, and I am still collaborating with Dan 40 years later in our studies of DNA.

Alec Bangham sparked my interest in the origin of life. In 1961, Alec discovered that phospholipid assembles into vesicles now called liposomes. I spent a sabbatical in Alec's lab at the Animal Physiology Institute in the tiny village of Babraham, a few miles south of Cambridge, England. One day while motoring down to London in Alec's Morris Mini, we came up with the following question: What lipid-like molecules were present on the prebiotic Earth that could assemble into the membranes required for the origin of cellular life? No one at the time had the slightest idea, but now, 30 years later, we can at least make an educated guess.

Tony Crofts hosted my first sabbatical leave in Bristol, England. Tony and I visited Alec to learn more about preparing liposomes, because we wanted to determine whether lipid vesicles could maintain pH gradients. Our study was inspired by a new idea called chemiosmosis that Peter Mitchell had published a few years earlier in *Nature*, so we also visited Peter, who had set up his own laboratory in a vast manor house just outside Bodmin. This was my first introduction to the extraordinary scientist whose ideas changed the way we think about bioenergetics.

Harold Morowitz and Franklin Harold both inspired me to think about how energy is captured by living cells today, and by extrapolation to life four billion years ago. Harold Morowitz wrote *Beginnings of Cellular Life*, the first book to focus on cellular compartments as an essential component of the earliest forms of life. Both Harolds wrote books on bioenergetics, describing with precision and clarity how energy flows through all of life.

John Oro was a true pioneer in advancing studies of life's beginnings. John discovered how adenine can be synthesized by the spontaneous polymerization of five HCN molecules, and was first to suggest that comets played an important role in delivering organic compounds to early Earth. John and I independently and simultaneously decided that we should add lipids to the list of prebiotic compounds. With our students as coauthors, we published separate papers describing how they can be synthesized in simulated prebiotic conditions, and then we coauthored a paper in 1980 that set the stage for involving lipids in origins of life studies.

Andrew Pohorille also studies lipids, particularly lipids in membranes and their dynamic motions. Andrew was kind enough to provide the computer images of lipid bilayers for this book.

Doron Lancet hosted my sabbatical at the Weizmann Institute in Rehovot, Israel, where he introduced me to an entirely new way to think about life's beginnings. With his student Daniel Segre, now at Boston University, we wrote a speculative paper entitled "The Lipid World" in which we imagined a living system with all the attributes of life, but no polymers.

Bob Hazen and Bernd Simoneit are geochemists who opened my eyes to the possible roles that mineral surfaces could play in the prebiotic chemical reactions required for the first forms of life. Bob wrote a wonderful book with the title *Genesis: The Scientific Quest for Life's Origin*, which was a constant inspiration to me as I thought about how to write this book.

Michael Mautner and Bob Leonard, Kiwi colleagues in Christchurch, hosted my visits to New Zealand. In one of my papers with Michael, we tested whether the soluble components of carbonaceous meteorites could provide a nutrient source for microbial and plant life. The amazing result is that they can! Bacteria and plants readily use atoms and molecules extracted from a meteorite that predates the age of the Earth.

Jerry Joyce and Jack Szostak are superb scientists who recognized early on that ribozyme catalysts in a cellular compartment are potential stepping stones toward the origin of life. Jack and Jerry independently showed how a system of enzymes and RNA could evolve, and Jack is now leading the way toward establishing a laboratory version of artificial life.

Then there are the authors of papers I have read with pleasure, learning from each. We regularly meet at Gordon Conferences, Astrobiology meetings, and ISSOL Congresses all over the world. Here they are, the colleagues who inspired many of the ideas that I discuss in *First Life*: Jeff Bada, John Baross, Steve Benner, Sherwood Chang, John Cronin, David Desmarais, Pascale Ehrenfreund, Jim Ferris, Nick Hud, Bill Irvine, Noam Lahav, Antonio Lazcano, Luigi Luisi, Lynn Margulis, Marie Christine Maurel, Stanley Miller, Leslie Orgel, Sandra Pizzarello, Carl Sagan, Bill Schopf, Steen Rasmussen, Jack Szostak, David Usher, Art Weber, and Richard Zare. Sadly, John C., Stanley, Leslie, and Carl are no longer with us in person, but their ideas are alive and will continue to influence our thinking for years to come.

I reserve particular thanks for Gail Fleischaker. In the 1980s, Gail was a graduate student working with Lynn Margulis, and her field of study was in the realm of philosophy of science. Gail was among the first to explore the idea introduced by Humberto Maturana and Francisco Varela, that life was unique in being autopoetic (self-initiating and self-sustaining) and that any definition of life must include bounded compartments. Gail and I had the pleasure of coediting a previous book called *Origins of Life: The Central Concepts*.

I also want to acknowledge the wonderful graduate students and postdoctoral associates who were brave enough to spend a few years in my lab. Why brave? Well, most grad students and post-docs expect to get a job when they complete their research, but there are virtually no jobs with the title Origins of Life Researcher. Despite this, all of them are gainfully employed, mostly in the academic world, but a few in industry and government labs. From A to Z, my special thanks go to Charles Apel, Gail Barchfeld, Ajoy Chakrabarti, Will Hargreaves, Dmitri Kirpotin, Sarah Maurer, Pierre-Alain Monnard, Trishool Namani, Felix Olasagasti, Sudha Rajamani, Sara Singaram, Roscoe Stribling, Sasha Volkov, and Helmut Zepik.

Not much happens in science without funding from research grants, and virtually all of my research described here was supported by grants from the NASA Exobiology and Astrobiology programs. And not much happens in publishing without agents and editors. Carol Roth found a home for this book at UC Press, and Blake Edgar shepherded the rough draft manuscript through a two-year process, gently guiding the writing task in helpful directions. I thank Kevin Plaxco, Jack Farmer, and Bob Hazen for their careful reading of the first draft and their kind comments.

Finally, I want to give my deepest thanks to my beautiful and smart wife, Ólöf Einarsdóttir, and our two children Ásta and Stella Davidsdóttír. Ólöf taught me all I know about Iceland and cytochrome oxidase, and Ásta and Stella taught me how great it is to be a father again.

INTRODUCTION

It was early evening in Baja California. The sun had disappeared behind the ragged mountains to the west, leaving the tiny bay of Puerto Escondido in darkness. My two kids were trying to get to sleep, and I could see the shadow of their mother moving in the tent, backlit by a candle. Several young graduate students, who had been noisily diving all day, carrying out the photography and tedious collecting of field research, were now asleep. A faint glow emanated from the quiet water of the bay. As I walked across the still warm beach, thin lines of phosphorescence flashed in the dark where wavelets lapped onto the sand. I waded into the warm Sea of Cortez, adjusted my diving mask and snorkel, and swam slowly away from the beach.

Peering down into the dark water, I was stunned. Constellations flashed against the glass of my mask. My arms glowed with a blue sparkling luminescence as they moved through the water. Galaxies of light swirled off my fingertips. I had to laugh, stopped swimming, and turned to float on my back. It was a moonless night, and the Milky Way filled the sky. An extraordinary sense of connection swept through me as I saw the starlight of our galaxy reflecting off the water. The photons had traveled hundreds, even thousands, of light-years, only to mix with the tiny flashing lights from the living organisms that filled the bay.

This sense of being one with the universe is the rarest human experience, and I am privileged to have felt it that night. Over the next 30 years, I found a way to study this connection as a scientist. Could there really be a connection between stars and life? Astrologers have always thought so, of course, but astronomers know better. At least they thought they did, until the birth of a new scientific discipline in 1996, when scientists at the Johnson Space Center in Houston, Texas, made a startling claim. They had discovered fossil microorganisms in a meteorite that was indisputably a chunk of the surface of Mars, sent sailing into space at escape velocity by the impact of a small asteroid on the Martian surface. I was with my family in Iceland when the news broke, and knowing my interest in such matters, my wife's sisters immediately called to tell me to check the front page. I hurried into downtown Reykjavik to find a copy of the local newspaper. There on the front page was a photo of President Bill Clinton talking about a stunning discovery made by people I knew at the Johnson Space Center and at Stanford University. The newspaper was in Icelandic, of course, so I used a dictionary to give myself a crash course in my wife's language and finally understood why everyone was so excited: Living organisms might have once flourished on a planet other than our own!

The excitement generated by this revelation was not lost on NASA administrators, who announced significant funding for a new research program called Astrobiology. That is how a seemingly impossible connection was made between astronomy and biology—by taking pieces of those two words, combining them into a new word, and most importantly, providing the research dollars in a competition that was certain to attract the finest scientific talent.

Two goals of astrobiology are to discover how life originated on our planet and to determine whether life exists beyond Earth. Because life today is so much a phenomenon of chemistry, it has been mostly chemists like me who are attracted to the question of how life began. We chemists perceive the origin of life as a chemical process. When the first microscopic organisms began to grow and reproduce on early Earth, chemical reactions associated with growth, metabolism, and replication were central to much of what we consider being alive. But how did that chemistry begin?

I believe the answer will be found in the realm of physics, and more specifically biophysics, defined as the physical processes that we now associate with the living state. The chemistry of life only became possible after certain physical processes permitted specific chemical reactions to occur. Life can emerge where physics and chemistry intersect.

EMERGENCE AND LIFE'S BEGINNING

In the previous sentence I used the word "emerge," which increasingly influences how scientists think about the origins of life. In common usage, "emergence" is an unexpected happening, as in an emergency. "Emergence" is now being used in science to connote the process by which a physical or chemical system becomes more complex under the influence of energy. There is a certain mysterious quality to the word's use in this regard because the emergent property is typically unexpected and cannot be predicted. Emergence is the opposite of reductionism, in which everything is believed to be explainable by understanding ever simpler components of a system. Reductionism, however, cannot account for the fact that under certain conditions, systems become increasingly and unpredictably complex.

The reality of emergent phenomena was first demonstrated by attempts to predict weather with mathematical models. Weather emerges when a source of energy (sunlight) interacts with vast masses of gas (the atmosphere) and water (the ocean). Equations were formulated that worked for a while but were then increasingly disturbed by extremely small variations in the numerical inputs so that the outcomes became unpredictable. Out of this discrepancy between mathematics and the real world grew an astonishing new idea called chaos theory, which held that certain aspects of physical reality are governed by processes that cannot be precisely described by a set of equations. When energy interacts with matter under certain conditions, we can confidently predict that *something* will happen, but we cannot always predict where it will happen or what it will be. That *something* that happens is what we refer to as emergence.

Because the central theme of this book is that living systems first emerged through the action of physical laws that, in turn, permitted certain chemical reactions to occur, the difference between physics and chemistry should be made clear. Chemical reactions are those processes by which changes occur in the electronic structure of atoms and molecules such that compounds with new properties are produced. Chemical reactions also alter the energy content of molecules. For example, when you light a match, the compounds in the match head react to release chemical energy as heat and light. Water and carbon dioxide, with greatly reduced chemical energy content, are by-products of that reaction. In contrast, a physical process typically changes the energy content of a system, but does not alter the electronic structure of its components. If you add energy to a soap solution by blowing air through a straw, the energy arranges soap molecules into transient structures called bubbles, but causes no permanent change in the properties of those soap molecules. One of the main arguments I will make in this book is that structures resembling microscopic soap bubbles were an absolute requirement for life to begin, as essential to the process as the assembly of genes and proteins.

DEFINING LIFE

An introduction to a book about the origins of life should at least try to define "life," but the fact is that no definition is generally accepted by biologists yet. Even the simplest microorganisms are extraordinarily complex, and dictionary-style definitions do not seem to encompass such complexity. However, in the next few years I think it is likely that someone will claim to have fabricated artificial life in the laboratory. To make that claim, he or she will need a satisfactory definition of life. Because life is a complex phenomenon, maybe the best we can do is to state a minimal set of properties that, taken together, exclude anything that is not alive. Here are the properties that I would include:

- Life is an evolving system of polymers synthesized by chemical reactions (metabolism) that take place in membrane-bounded compartments called cells.
- Polymers are very long molecules composed of subunits called monomers. The primary polymers of life are nucleic acids and proteins, often called biopolymers.

- Biopolymers are synthesized by linking together monomers—amino acids and nucleotides—using energy available in the environment. Polymer synthesis is the fundamental process leading to growth of a living system.

- Nucleic acids have a unique ability to store and transmit genetic information. Proteins called enzymes have a unique ability to act as catalysts that increase the rates of metabolic reactions.

- The genetic and catalytic polymers are in a cyclic feedback system in which information in the genetic polymers is used to direct the synthesis of the catalytic polymers, and the catalytic polymers take part in the synthesis of the genetic polymers.

- During growth, the cyclic system of polymers reproduces itself, and the cellular compartment divides.

- Reproduction is not perfect, so variations arise, resulting in differences between cells in a population.

- Because different cells have varying capacities to grow and survive in a given environment, individual cells undergo selection according to their ability to compete for nutrients and energy. As a result, populations of cells have the capacity for evolution.

THE HYPOTHESIS

With this working description of life, we can ask how such a complex system of molecules could have emerged on the surface of a sterile planet like early Earth. Many previous proposals about the origins of life are general and lack details, such as Aleksandr Oparin's proposal in 1924 that life began as microscopic assemblies of polymers. At the other end of the spectrum are highly specific ideas—for instance, that life began as a self-replicating RNA molecule, or as a thin film on the surface of an iron sulfide mineral called pyrite. We test such ideas in the laboratory by fabricating what are called "plausible simulations" of prebiotic conditions. A classic example is Stanley Miller's experiment reported in 1953, in which a gas mixture was exposed to an electrical discharge, with the remarkable result that amino acids were synthesized.

What I will propose in this book is an integrated set of ideas and themes that suggest a new way to think about the origins of life. The primary themes are cycles, compartments, and combinatorial chemistry. Taken together, these themes suggest a novel approach that is guided by the principle of sufficient complexity, in which the origin of life is understood as an emergent phenomenon that occurs when water, mineral surfaces, and atmospheric gases interact with organic compounds and a source of energy. If we are to understand how something as complex as life can begin, we need to make a laboratory simulation sufficiently complex to accommodate cycles, compartments, and combinatorial chemistry. Each chapter of this book presents aspects of these central themes, like pieces of a puzzle, and in Chapter 14, I will integrate the pieces into a descriptive scenario, essentially a scientific hypothesis with testable predictions. Because science works best when there are alternative hypotheses that can be discriminated by critical experiments, these alternative ideas will also be described and evaluated. Finally, science necessarily progresses by filling in gaps with essential knowledge, so I will also try to make explicit the gaps I perceive as a scientist working on the problem of life's origins.

In order to give a sense of how the hypothesis developed, I want to outline the narrative that will be presented. The first forms of life assembled from organic compounds composed of carbon, hydrogen, oxygen, nitrogen, phosphorus, and sulfur. These elements were synthesized in stars and then delivered to planets like Earth when solar systems formed in vast clouds of dust and gas. Early Earth had oceans, volcanic land masses, and an atmosphere of carbon dioxide and nitrogen gas. The most plausible site for the origin of life was not the open ocean or dry land. Instead, there is reason to think that the most conducive conditions for life to begin were places where liquid water and the early atmosphere encountered mineral surfaces such as volcanic rocks. We call the contact between liquid, solid, and gas an interface. As you will see, interfaces have special properties because they allow certain essential processes to occur that do not happen anywhere else: wet-dry cycles, concentration and dilution, formation of compartments, and combinatorial chemistry.

· *Cycles:* The local environment probably resembled pools in volcanic sites where hot water constantly goes through cycles of wetting and drying. We call this a fluctuating environment. The pools contained complex mixtures of dilute organic compounds from a variety of sources, including extraterrestrial material delivered during the last stages of Earth's formation, and other compounds produced by chemical reactions associated with volcanoes and atmospheric reactions. Because of their fluctuating environment, the compounds underwent continuous cycles in which they were dried and concentrated, then diluted upon rewetting.

· *Compartments:* During the drying cycle, the dilute mixtures formed very thin films on mineral surfaces—a process that is necessary for chemical reactions to occur. Not only did the compounds react with one another under these conditions, but the products of the reactions became encapsulated in microscopic compartments by membranes that self-assembled from soaplike organic compounds called amphiphiles. This process resulted in vast numbers of protocells that appeared all over early Earth, wherever water solutions were undergoing wet-dry cycles in volcanic environments similar to today's Hawaii or Iceland.

· *Combinatorial chemistry:* The protocells represented compartmented systems of molecules, each different in composition from the next, and each representing a kind of microscopic natural experiment. Most of the protocells remained inert, but a few happened to contain a mixture that could be driven toward greater complexity by capturing energy and smaller molecules from outside the encapsulated volume. As the smaller molecules were transported into the internal compartment, energy was used to link them into long chains.

Today we call the smaller molecules monomers and the long chains polymers. Examples of monomers today are amino acids and nucleotides, which form polymers called proteins and nucleic acids (DNA and RNA). The polymers have emergent properties that are far beyond what the monomers can do. Most important is that both kinds of polymers can act as catalysts, and one of them—nucleic acids—can carry and transmit genetic information in specific sequences of the monomers that compose it.

Life began when one or a few of the immense numbers of protocells found a way not only to grow but also to incorporate a cycle involving catalytic functions and genetic information. According to this hypothesis, cells, not molecules, were the first forms of life.

It is the task of scientists interested in the origins of life to discover how all this could have taken place in the prebiotic environment. In the chapters to follow, I will trace the trajectory of this research within the purview of a new scientific discipline called astrobiology, which studies the origin and evolution of life on Earth as part of a universal process involving the birth and death of stars, the formation of planets, the interfaces between minerals, water, and the atmosphere, and the physics and chemistry of carbon compounds. I will report on the progress we are making toward synthesizing life in the laboratory, and whether life began elsewhere in our solar system. I also want to include personal experiences of the people who are doing the work. Science is commonly viewed as just a body of knowledge, but for those engaged in research it is much more than that because the process is embedded in a distinct human culture that is rarely revealed. The fact is that only 1 in 1,000 people in Europe and the United States makes a living doing science, and only one in a million works as an astrobiologist. This is a very rarefied atmosphere, as rarefied as the number of billionaires in the human race, only about 1,000 people.

Scientists rarely become billionaires, but they have discovered a vast source of a different kind of wealth. That wealth is the treasure of discovery, the endless pleasure of asking questions and seeking answers—and it is there for the taking. During their careers, scientists ask hundreds of questions and try out thousands of possible answers, or hypotheses. Asking good questions and developing testable hypotheses are the two essential creative acts that scientists must engage in. Most of our hypotheses are wrong, but occasionally we discover an answer that survives critical experimental tests and adds a new dimension to the map of reality we call science. When this happens, the experience is so intensely satisfying that we willingly spend our lives pursuing it.

1

A FIREBALL OVER AUSTRALIA

In the summer of 1981, a small black stone wrapped in aluminum foil changed the course of my life. About the size of a marble and indistinguishable from any other rock that might be found on a beach, this stone had traveled from southeastern Australia to NASA Ames Research Center in Mountain View, California, where researcher Sherwood Chang showed me the specimen and gave me a small sample to test. But the stone had traveled even farther. It was a piece of a meteor that had lit up the night sky over the town of Murchison, Australia, in September 1969. The fall began with a bright orange fireball and rolling thunder, followed minutes later by a shower of black stones strewn over five square miles. During the next few weeks, townspeople and scientists collected more than 100 kilograms of meteorites ranging in size from marbles to bricks.

Why would a marble-sized rock be so significant that it can change one's life? The reason is that before traveling 7,000 miles from Australia to California, this tiny bit of rock traveled to Earth from the asteroid belt between Mars and Jupiter, a distance of 250 million miles, or even billions of miles if we take into account all of the rock's orbits around the sun before reaching Earth. The rock was produced when a smaller asteroid happened to collide with a larger one, knocking off fragments of the surface and leaving behind a crater resembling those that we can see in photographs of asteroid surfaces (Figure 1).

I held in my hand a genuine rock from outer space. But there was more to this meteorite than just the mineral content typical of other stony meteorites. When the original boulder-sized object exploded over Murchison, the surfaces of the smaller fragments

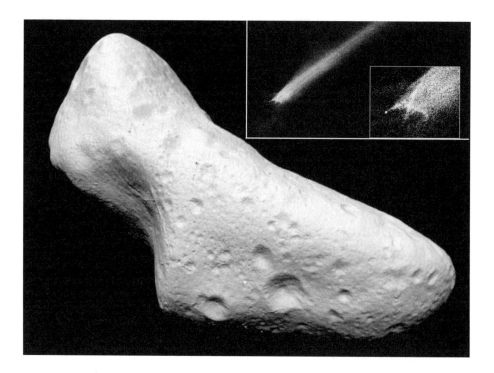

FIGURE 1

The asteroid Eros shows numerous craters resulting from collisions with smaller objects in the asteroid belt. Each impact has the potential to produce fragments of the asteroid surface that can reach Earth as meteorites. The inset shows an actual collision between two asteroids that happened to be photographed, and the shower of fragments can be clearly seen.

were heated by atmospheric friction to white-hot temperatures. In a few seconds, the friction slowed the fragments from an initial velocity of 20 kilometers per second, and they finally fell to the ground at the same speed they would reach if they were dropped from an airplane. The first stones to be discovered were still emitting a smoky smell from their hot surfaces, a distinctive aroma that I would later notice while extracting organic compounds from the meteorite. Whenever I give talks about this work, I evaporate a drop of Murchison extract into a wine glass and pass it around for audience members to sniff, joking that 4,570,000,000 BCE was a very good year.

That age, 4.57 billion years, is the reason why certain meteorites have changed scientific lives. The Murchison meteorite belongs to a relatively rare group of meteorites called carbonaceous chondrites. Their aroma is produced by organic compounds older than Earth itself, some of which were present in the vast molecular cloud of interstellar dust and gas that gave rise to our solar system 4.57 billion years ago. Most of the organic material—nearly 2% of the total mass of a typical Murchison sample—is in the form of a tarlike polymer called kerogen, but there are also hundreds of different compounds that sound like a chemist's laboratory: oily hydrocarbons, fluorescent polycyclic aromatic hydrocarbons (PAHs), organic acids, alcohols, ketones, ureas, purines, simple sugars, phosphonates, sulfonates, and the list goes on. Where did all this stuff come from? Did it have anything to do with the origin of life?

THE MURCHISON METEORITE AND SELF-ASSEMBLY

With a sample of a carbonaceous meteorite in hand, I was ready to do an experiment I had been dreaming about. Ten years earlier, shortly after the Murchison event, Keith Kvenvolden and a group of researchers at NASA Ames had analyzed a sample of the meteorite and convincingly demonstrated that amino acids, one of the essential organic compounds composing all life on Earth, were present in the meteorite. And these were not just the kinds of amino acids found on Earth (which might have been contamination), but more than 70 other kinds that were clearly alien to biology as we know it. This study, and many that followed, established that amino acids, the fundamental building blocks of proteins, can be synthesized by a nonbiological process. From this, it seems reasonable to think that amino acids, at least, would have been available on prebiotic Earth.

I had spent much of my earlier research career studying lipids, which, along with proteins, nucleic acids, and carbohydrates, represent the four major kinds of molecules that compose living organisms. "Lipid" is a catch-all word for compounds like fat, cholesterol, and lecithin that are soluble in organic solvents. In earlier research I had extracted triglycerides (fat) from the livers of rats, phospholipids such as lecithin from egg yolks, and chlorophyll from spinach leaves. All of these procedures used an organic solvent mixture of chloroform and methanol to dissolve the lipids, and I wanted to try the same thing with the Murchison material. The surface of the meteorite certainly had surface contamination from being exposed to the laboratory atmosphere, so I broke it into smaller pieces and carefully obtained an interior sample weighing about 1 gram. Then I ground the sample in a clean mortar and pestle with a mixture of chloroform and methanol as the solvent, and decanted the clear solvent from the heavier black mineral powder. The chloroform solvent had a yellow tint, which meant that it had dissolved some of the organic material in the meteorite. I dried a drop of the solution on a microscope slide, added water, and then examined it at $400\times$ magnification. It was an extraordinary sight. Lipidlike molecules had been extracted from the meteorite and were assembling into cell-sized membranous vesicles resembling microscopic soap bubbles (Figure 2). Could it be that similar compartments were present when the first liquid water appeared on Earth more than four billion years ago? Maybe, just maybe, if we studied the Murchison meteorite we might know what kinds of molecules made up the membranous boundaries of the first cellular life.

But a huge question remained: Where did the stuff come from? For that matter, where does anything come from? To set the stage for telling that story, let's begin with stars, where everything begins.

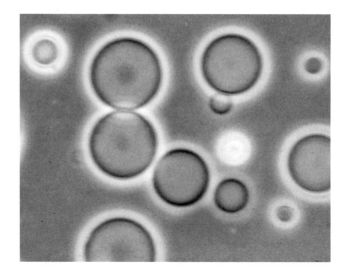

WHERE DOES EVERYTHING COME FROM? THE LIFE AND DEATH OF STARS

Forty years ago, when a boulder-sized meteorite blazed through the skies above Murchison, Australia, we had only a few speculations about when our universe began and how galaxies, stars, and planets come to be. Now, within a single lifetime, we have definite answers to fundamental questions that have been pondered throughout recorded history. The first hint of such an answer was put forward in 1946 by Fred Hoyle, at the time a young, brash British astronomer. Born in 1915 in Yorkshire, England, Hoyle displayed a creative genius that is a wonderful example of how ideas become grist for the mill of science, and how bad ideas disappear into dust, while the rare gems of good ideas survive scientific grinding to become touchstones for future generations of scientists. Fred Hoyle was full of ideas and was bold enough to publish them. Here are some examples:

· The fossil archaeopteryx (a small, feathered, birdlike dinosaur) in the British Museum of Natural History is a fake.

· Vast molecular clouds in outer space are loaded with microorganisms which brought the first forms of life to Earth.

· Viruses that cause flu epidemics are brought to Earth when it passes through the tails of comets.

· All the carbon required for life to exist is synthesized in the interior of stars.

· The universe has no beginning or end. The idea that the universe has a beginning is nonsense, and it deserves a silly name: the Big Bang.

· The requirements for the synthesis of carbon are so precise that life could not have accidentally arisen. There must have been an intelligence at work to make it happen.

Of these ideas, only one has survived the experimental and theoretical tests that are characteristic of the scientific enterprise. It is now the consensus that all the carbon circulating throughout the universe, including every carbon atom in your body, was synthesized in extremely hot interiors of dying stars (100 million degrees!) and then blasted out into space when the star reached the end of its lifetime in a nova or supernova explosion. To understand this process, we need to recall a little high school chemistry. All matter is composed of atoms, and all atoms have a tiny nucleus composed of elemental particles called protons and neutrons, which are surrounded by orbital clouds of much lighter electrons. (Protons and neutrons are approximately 1,800 times more massive than electrons.) But in stars, the temperature is so high that the electrons cannot be held in place, so basically, stars like our sun are composed of a gas of naked atomic nuclei, mostly in the form of hydrogen and helium. Hydrogen is the lightest element, with a single proton and no neutrons in its nucleus. Helium is the second lightest element, with two protons and two neutrons in its nucleus, and is the product of the initial hydrogen fusion reaction that makes stars shine. If we could somehow grab a 1-gram sample of the universe and use it to fill a balloon on Earth, the balloon would float away because the visible material in the universe is mostly hydrogen and helium.

What Hoyle brilliantly realized is that, at sufficiently high temperatures, two helium nuclei can fuse to form a nucleus of the lightest metallic element, beryllium, which then fuses with a third helium nucleus to produce carbon. Because three helium nuclei combine to make one carbon, this reaction is called the triple alpha process. (Helium nuclei are also called alpha particles when they are emitted from a radioactive element.)

Hoyle had his revelation about the origin of carbon in 1946, but earlier theoretical models had already shown that if carbon is somehow made available, nitrogen and oxygen can be formed in a process called the carbon-nitrogen-oxygen (CNO) cycle, which turns out to be the primary source of fusion energy in large, hot stars on their way to oblivion as novas and supernovas. Descriptions of the CNO cycle were independently published by Carl von Weizsäcker and Hans Bethe in 1938 and 1939. They did not know how the carbon was made, and this is where Hoyle filled in a significant gap in our knowledge a few years later.

Hoyle published his idea in 1946 but did not include a mathematical analysis, although he hinted at it. In 1957, Hoyle joined William Fowler, and Margaret and Geoffrey Burbidge at Cal Tech to publish an article in *Reviews of Modern Physics,* which became a classic. The article was a brilliant analysis of nucleosynthesis of elements in stellar interiors, and should have won a Nobel Prize for someone. It did, but not for Hoyle. Discovering the treasures hidden in the scientific landscape is a chancy business, but getting credit for your discoveries is even chancier. The Nobel Prize in 1983 went to Fowler, who certainly deserved it for his many contributions, and the prize was shared with Subrahmanyan Chandrasekhar, who studied stellar evolution. Sir Fred had to be satisfied with a knighthood.

To sum up, we can now account for all the major elements of life in terms of nuclear reactions in stars, a process called stellar nucleosynthesis. The atoms of carbon, nitrogen, oxygen, sulfur, and phosphorus that comprise all life on Earth were once in the center of stars more massive than our sun, forged at temperatures hotter than any hydrogen bomb. And what about hydrogen? Even more astonishing, most hydrogen atoms are as old as the universe, which somehow burst into existence 13.7 billion years ago when time began. As living organisms, we are not in any way separate from the rest of the universe. Instead, we simply borrow a tiny fraction of its atoms for a few years and incorporate them into the patterns of life. The hydrogen and oxygen atoms are in the water that flows through our cells, and the carbon, hydrogen, oxygen, sulfur, and phosphorus are linked together in the proteins, lipids, and nucleic acids that are the stuff of life. This is why we call them biogenic elements.

STEADY STATE AND BIG BANG FACE OFF

We have a fair understanding of how the elements of life are made, but what about the universe itself? This story has been told many times, but it is such a good story that I can't resist telling it again. This means that I must first tell you about George Gamow, who has been part of my life for as long as I can remember. As a teenager, all I knew about the person whose name I mispronounced (it's "Gamoff," not "Gamau") was that I could actually understand what he wrote about cosmology. A tattered paperback copy of his book *One . . . Two . . . Three . . . Infinity!* has a place of honor on my library shelf.

Gamow and Hoyle, both energetic personalities, full of good and bad ideas, came to loggerheads over one of the great questions of all time: Did the universe have a beginning? One answer had already been suggested by Monsignor Georges Lemaître, a remarkable Catholic priest whose lifetime (1894–1966) corresponded closely with Gamow's (1904–1968). In 1931, Lemaître published a paper in *Nature* in which he proposed that the universe was expanding, and must therefore have begun as a "primitive atom." This was not just an idea but was based on a substantial mathematical foundation. When Albert Einstein met Lemaître a few years later, he told the priest, "Your calculations are correct, but your physics is abominable."

Lemaître was vindicated two years later when Edwin Hubble presented direct evidence that the light from distant galaxies had a longer wavelength than nearby galaxies, a phenomenon called the red shift. Hubble's observation led to a revelation about our universe, so it deserves a bit of explanation to underline its significance. The simplest way to understand the red shift is by analogy to sound produced by a vibrating structure. Sound travels through air as a vibration of a certain frequency, with lower tones having lower frequencies measured as vibrations per second. For instance, the note A in the middle of a piano keyboard vibrates 440 times per second. Light also has wavelike properties, but its frequency is a trillion times that of sound. The thing to keep in mind is that red light has a lower frequency than blue light.

Most of us have stood near a road when a car passes with its horn blaring. We hear a higher horn tone when the car is approaching, then lower in tone after it passes. This is called the Doppler effect, in honor of Christian Doppler, who first proposed an explanation in 1842. Now imagine that the car passes by at nearly the speed of light, and that we are seeing the headlights rather than listening to the horn. As the car approaches, the headlights would look blue, and as it speeds away from us, they would look redder. This is the effect that Hubble observed, now called the red shift. The most plausible explanation is that galaxies are moving away from us and that the farthest objects have the greatest apparent velocities, some in fact approaching the speed of light.

Gamow loved the idea of an expanding universe. In 1948, he published a superb paper with his student Ralph Alpher entitled "The Origin of Chemical Elements." The main point of the paper is that hydrogen and helium compose most of the matter of the universe, and they are present in a certain ratio: 92% hydrogen to 8% helium, in terms of the number of atoms. Gamow also included in the paper an important prediction: When it popped into existence, the universe must have been very hot, hotter than the hottest stars today. But with the passage of time, and the expansion of the universe, the cosmic temperature must decrease, just as a compressed gas cools off when it is allowed to expand. Gamow predicted that if we could somehow listen to the universe, we should still be able to "hear" this energy as a kind of low rumble of radio waves.

When he submitted the paper for publication, Gamow could not resist adding the name of his friend and colleague Hans Bethe, who had nothing to do with the work. Gamow expected his joke to be caught before publication, but no one noticed. The paper was published in *Physical Reviews*, appropriately on April 1, 1948, and the authors listed were Alpher, Bethe, and Gamow. If you don't get Gamow's joke, don't feel bad, because his editor and peer reviewers—all professional physicists—missed it, too. The first three letters of the Greek alphabet are *alpha, beta,* and *gamma,* which also refer to the primary particles released by radioactive decay.

Hoyle did not agree that the universe had a beginning. In 1948, he published an alternative hypothesis with his colleagues Hermann Bondi and Thomas Gold, which came to be known as the steady state theory. Perhaps the universe was expanding, they reasoned, but instead of beginning as Lemaître's primitive atom, matter was continuously created out of a hypothetical source of energy to replace the matter lost to expansion, thereby maintaining the steady state we seem to observe today. Hoyle ridiculed Gamow's ideas, referring to them in jest during a radio show as the "Big Bang." The name stuck.

The opposing ideas of Hoyle and Gamow are an example of science at its best, when two alternative hypotheses are available for testing. The influential philosopher of science, Karl Popper, proposed that even the most creative ideas cannot be classified as science unless they are "falsifiable," or testable by experiment or observation. In 1962, Thomas Kuhn wrote a book called *The Structure of Scientific Revolutions*, in which he proposed that on rare, exciting occasions, the accumulation of such evidence will cause one paradigm, or consensus view, to crumble while an alternative paradigm rises to take its place. The goal, then, is to find a critical experiment, usually in the form of a prediction, that will permit us to choose between alternative hypotheses. Sometimes this happens by an accidental observation that we call serendipity, after the folk tale about the three princes of Serendip (an old name for Sri Lanka) who constantly found things by accident that happily fulfilled a future need. In the case of the universe, it was two Bell Labs scientists who stumbled across the answer, and a Princeton physicist who recognized the significance of their discovery.

The prediction had already been made in Gamow's 1948 paper, which was also independently proposed by Yakov Zel'dovich working in Russia, but not much attention was paid to it until Robert Dicke at Princeton had a similar idea. In 1964, two other faculty members at Princeton, David Wilkinson and Peter Roll, began to construct an antenna and amplifier specifically designed to look for cosmic background radiation, which was predicted to exist as a consequence of the explosive birth of the universe. That is, if the Big Bang idea was correct, the energy level of the universe was expected to have cooled to a temperature just a few degrees above absolute zero, a temperature that could be detected as a specific radio frequency in the microwave region of the radio spectrum. By a remarkable coincidence, Arno Penzias and Robert Wilson at the Bell Labs were testing a similar instrument that was originally used for bouncing radio waves off satellites. It worked, but there was a strange problem. No matter what direction in the sky they pointed the long metal horn that collected radio waves, it picked up a microwave radiation equivalent to a temperature just a few degrees above absolute zero. This was crazy! Penzias and Wilson had expected to see a certain amount of noise from a variety of sources, and had adjusted their instrument so that ordinary radio waves and errant radar pulses were tuned out. They cooled the receiver with liquid helium, and even cleaned out pigeon droppings that had been deposited in the metal horn over the years. Then they happened to hear about a paper being written (but not yet published) by Dicke and his associates in which it was made clear that a suitable detector should be able to hear the cosmic microwave background, now abbreviated CMB. Penzias and Wilson had inadvertently constructed just such an instrument, and they suddenly realized the importance of their result. Penzias called Dicke and invited him to drive 40 miles from Princeton to the Crawford Hill Bell Labs and literally listen to the CMB, which they could hear as a kind of rumbling static in their headphones.

Everything fit. In a nice example of scientific generosity, in 1965 the two groups published joint papers in the *Astrophysical Journal*. The first was authored by Dicke, Peebles, Roll and Wilkinson, and outlined the theoretical background of the prediction. The second paper described the detection of the CMB by Penzias and Wilson, for which they were awarded the Nobel Prize in 1978.

With an actual measurement of the predicted CMB, Hoyle's steady state hypothesis had been refuted, and the Big Bang hypothesis became an accepted theory. Gamow must have relished the shift in his favor, but he lived for only three more years. He spent his last years as a faculty member at the University of Colorado, dying in 1968 from liver disease. Gamow Tower at the Boulder campus is a memorial to his time there. But, like Sir Fred, Gamow was never honored with a Nobel Prize. Michio Kaku, a theoretical physicist best known for his contributions to string theory (the "theory of everything") and his radio show *Explorations,* once commented that Gamow's willingness to write children's books, like the one that had fascinated me as a teenager, probably influenced the Swedish committee members who choose Nobelists every year. Kaku wrote that "people could not take him seriously when he and his colleagues proposed that there should be a cosmic background radiation, which we now know to be one of the greatest discoveries of 20th-century physics."

HOYLE, WICKRAMASINGHE, AND PANSPERMIA

Hoyle lived on another 35 years. Although he was never able to match his earlier triumph of carbon nucleosynthesis, he certainly tried. Together with his colleague Chandra Wickramasinghe, now at Cardiff University in Wales, Hoyle coauthored a series of books and papers in which they speculated about how life could have begun on the Earth. The most widely accepted hypothesis is that life began as a chance event in which just the right mix of organic compounds was acted upon by an energy source so that growth and reproduction could occur. The earliest life would not resemble today's highly evolved version, but more likely it was a kind of scaffold that had the essential properties of life. The scaffold was then left behind when more efficient living systems evolved.

Hoyle and Wickramasinghe did not subscribe to this view. Instead, they elaborated an older idea championed in 1903 by Svante Arrhenius, the great Swedish chemist. Known as panspermia, this idea proposes that life exists everywhere in the universe and that life began on Earth when frozen extraterrestrial bacteria or spores, drifting as interstellar dust through the galaxy, happened to land here four billion years ago and found it to be habitable. Hoyle took it a step further when he claimed that this was still happening, that epidemics such as the flu pandemic of 1918 were actually caused by extraterrestrial organisms in the tails of comets.

I once met Wickramasinghe in 1986 at the Tidbinbilla radio telescope observatory near Canberra, Australia, and asked whether he and Hoyle really thought that interstellar space was infested with bacteria. He was quite certain of it, he said, noting that the infrared spectrum of interstellar dust closely matched that of dried, frozen bacteria. I mentioned that I was working with the astronomer Lou Allamandola at NASA Ames Research Center, who had demonstrated that the infrared spectrum could be reproduced by ordinary compounds called polycyclic aromatic hydrocarbons (PAHs). This seemed a much more plausible explanation than a galaxy full of bacteria. Wickramasinghe had a ready retort: "It is up to you to prove that they are not bacteria."

I have found that a few of my colleagues are not swayed by plausibility arguments, or Occam's Razor and the weight of evidence. In general, scientists are like investors, but instead of money, they invest time—limited to roughly 40 years of active research. Scientists are continuously making judgment calls to decide how to invest their time. They hope their investment will be profitable, not necessarily in monetary terms (that rarely happens) but rather in revealing significant new knowledge. A few scientists spend their lives seeking unusual explanations that others would immediately discard as implausible. Some of my colleagues avoid interacting with these mavericks, but I enjoy listening and reacting to their ideas. Most often the ideas turn out to be not just implausible but wrong. Yet once in a while, a wild idea is beautifully, wonderfully correct. George Gamow had one such idea, and later in this book, I will tell you about Peter Mitchell, another maverick whose implausible idea taught us how metabolic energy is captured in ATP in all life today.

STARS, DUST, AND SOLAR SYSTEMS

The current consensus regarding the time scale of events following the beginning of our universe represents an absolute revelation in the history of science, made possible only when we had a mix of data from radio astronomy and images from the orbiting Hubble Space Telescope that captures light from galaxies at the edge of the visible universe, light that is over 12 billion years old. Using this data, and applying well-established laws of physics, cosmologists can conceive internally consistent scenarios of events that occurred within the first second, minute, hour, and day after the Big Bang, and then the billions of years that followed. The most recent surprise is that the universe seems to consist mostly of "dark matter" and "dark energy" with ordinary matter comprising only a small fraction of its mass—just 4% by one estimate. Furthermore, the expansion is not slowing down, as might be expected, but instead is speeding up. We still have much to learn about the universe we live in, a universe so remarkable that its laws allow life to begin on a tiny planet circling an ordinary star, then to evolve into conscious human beings who can wonder where it all came from.

What in this model is relevant to the origin and distribution of life? One thing to realize is that there were no stars for the first 400 million years. The hydrogen fusion reactions necessary for stars to exist could only begin after the universe had cooled sufficiently for hydrogen to collapse by gravitational attraction into dense structures such as stars. Because the Big Bang distributed hydrogen unevenly, the first stars to form clumped together by the billions into immense galaxies. We call our own galaxy the Milky Way, and the word "galaxy" is in fact derived from the Greek and Latin words for "milk." If we could view our galaxy from above, we would see not only hundreds of billions of stars, but also dark clouds of dust throughout the spiral arms.

Where does the dust come from? When the first stars and galaxies appeared, close to 13 billion years ago, there was no dust, instead only hydrogen and helium were produced during the Big Bang. There were no planets or solar systems because elements heavier than hydrogen and helium were still buried in the hot nuclear gas composing the first generation of stars. How did they escape? The answer to this question comes from yet another revelation of modern astronomy. Even though stars seem unchanging during a human lifetime, on the immense time scale of the universe, stars burn for millions or billions of years and then die in spectacular explosions called novas or supernovas. This is how the heavier elements produced in the first generation of stars escaped into space, finally becoming available to form new stars and solar systems.

In 1054 CE, Chinese astronomers recorded the appearance of a new star as one of the first observations of a supernova, so bright that it shone even during the day. It disappeared after a few weeks, but 700 years later, Charles Messier discovered something strange in the night sky: not a sharply defined star, not a moving comet, but instead a stationary blur of light. Messier had spent his life using a primitive telescope and made a catalog of anything he saw that was not a star or a comet. This faint, blurry object was the first to be described in his catalog, so it is referred to as M1. The catalog was published in France in 1774, and 100 years later, telescopes had improved to the point that M1 could be seen as more than a faint blur. An early drawing of it showed a crablike structure, so M1 became known as the Crab Nebula.

Recent images taken by the Hubble orbiting telescope (visual light), the Spitzer orbiting telescope (infrared light), and the Chandra orbiting telescope (X-rays), show this amazing object in multiple colors. (See "Sources and Notes" for Chapter 1 at the end of this book for a collection of Crab Nebula images.) Just in the center you can see a small star, which is all that is left of the original star that was about 10 times more massive than our sun. Now it is an extremely dense ball of neutrons about the size of a small city, spinning 30 times per second and emitting a low-frequency "hum" that is detectable with a radio telescope. We refer to such stars as pulsars because of the pulsing radio waves produced as they spin. Pulsars are not smooth spheres, but instead have a region that sends radio frequency energy into the surrounding space, much as a lighthouse sends a beam of light out from a rotating lamp. If you would like to hear some of these literally unearthly stellar noises, search for "pulsar sounds" on the Internet.

The reddish glow of the Crab Nebula is due to dust particles that are illuminated by its central star. This stardust will later accumulate into vast clouds in the galaxy and give rise to new stars and solar systems, composed of ashes from the Crab and other exploded stars. To give some idea of the scale, the nebula is 6,500 light-years distant and 12 light-years across. Our entire solar system out to the orbit of Pluto is only 10 light-hours across, so approximately 10,000 of our solar systems lined up edge to edge would fit into the diameter of the Crab Nebula.

Supernova explosions are rare. In our galaxy, which contains 400 billion stars, a supernova occurs on average once every 50 years. They are produced by stars larger than our sun that are very hot and rapidly burn through their hydrogen fuel. In contrast, ordinary stars have lifetimes of several billion years, thousands of times longer than the stars that end up as supernovas. A consensus age of the universe is 13.7 billion years, which means that we are only just now seeing a generation or two of ordinary stars reach the end of their 5- to 10-billion-year lifetime. When stars like our sun deplete the hydrogen that fuels their fusion furnace, they swell into enormous, cooler red giants, then collapse to become white dwarfs. Our sun is middle aged, with perhaps five billion years to go, but when its red giant phase begins it will expand as far as the orbit of Mars and the four inner planets of our solar system will become baked cinders.

STARDUST AND THE ORIGIN OF SOLAR SYSTEMS

When a red giant collapses, more than half of the remaining mass is propelled into the interstellar medium, often producing beautiful structures called planetary nebulas that slowly expand from the central star. Observational astronomers seldom have a chance to exercise their imaginations, so whoever first sees such images appearing on his or her photographic plates (or computer screens these days) gets to say what they look like, something like a stellar Rorschach test. Examples include the Ant, the Eskimo, and the Cat's Eye nebulas.

These stunning images can be viewed in the Hubble Heritage website (see "Sources and Notes"), and only became possible after the Hubble Space Telescope (HST) was launched into orbit April 24, 1990. We are the first generation of human beings ever to see the true beauty of our galaxy and universe. The Hubble cost $2.5 billion to launch, two dollars per year for every US taxpayer, about what I spend in the morning for an espresso coffee. The return on this modest investment has been an immense expansion of our knowledge about the origin and evolution of the universe we live in. We can also see the structure of molecular clouds in our galaxy and get a glimpse of new stars and solar systems in the process of formation. We can see nearly to the edge of the universe, and the HST has captured perhaps 1,000 distant galaxies in a single image as it peered into the depths of space.

What do supernovas and planetary nebulas have to do with the origins of life? The answer lies in the composition of the material that is ejected during the explosive phase, and what happens to it over time. The basic composition is not very exotic: whatever was in the original star, but cooled down as it expands into space. This includes a mixture of elements in the gas phase, such as hydrogen, helium, nitrogen, oxygen, and sulfur; simple compounds like water and carbon dioxide; dust particles composed of silica, the same stuff as in sand; and metals like iron and nickel. If this sounds familiar, it should, because it is what Earth—and everything alive on Earth—is made of. Over millions of years, the dust slowly gathers into enormous clumps light-years in diameter, and these clumps give rise to new stars and solar systems.

One such interstellar cloud is called the Rosette Nebula, and magnificent images can be viewed in the Hubble Heritage website (see "Sources and Notes"). Dark dusty clouds are clearly outlined against the background, which is illuminated by a group of young stars that have burst into life. The pressure of their light is blowing away the surrounding dust so that a space has been cleared around the group, leaving a hole in the center of its rosette shape. The Rosette Nebula is about 50 light-years across, 10 times the distance from our solar system to the nearest star, Alpha Centauri. Although the Rosette Nebula is particularly beautiful, the formation of new stars and solar systems occurs throughout our galaxy, wherever nebular clouds of stellar ashes accumulate.

Five billion years ago, our own solar system emerged from a cloud resembling the Rosette Nebula and it is important for our story that we understand how it happened. This brings us to planet formation, and how Earth came to be the way it is. Let's go through the steps in chronological order:

1. Clouds of dust and gas ejected from dying stars gather into immense aggregates like the Rosette Nebula.

2. Turbulence within a cloud (or sometimes a shock wave from a nearby supernova) causes the local density to increase to the point that gravity takes over and produces dense aggregates of dust and gas called proplyds.

3. As the dust and gas collapse inward under the force of gravity, residual rotation within the proplyd speeds up, just as ice skaters spin faster when they pull in their arms.

4. The rotating cloud of dust and gas forms what is called a nebular disk, which has a much higher density in its center.

5. Material falls into the dense center ever more rapidly, and the heat generated begins to raise the temperature. The protostar begins to glow red.

6. The temperature at the center rises to 10 million degrees, and hydrogen fusion begins. The protostar has become a true star.

7. But in some cases, not all of the material in the nebular disk falls into the new star. Some of it can remain to form what is called a protoplanetary disk. This material has its own turbulence and rotating local densities, and undergoes smaller versions of gravitational collapse. The result is a series of planets circling the star, each with its own residual rotation.

8. In the case of our own solar system, in the first few million years after the sun reached its full status as a new star, its light and solar wind began to exert pressure on the remaining dust and gas of the disk. The material was slowly blown outward, leaving behind four rocky planets, including Earth. But beyond the orbit of Mars, the effects of pressure became negligible so that the planets now called Jupiter, Saturn, Uranus, and Neptune had time to gather the remaining material and become the gas giants we see today.

Although we can never see this process from beginning to end in a single lifetime, the reason the story is so convincing is that with the help of the Hubble Space Telescope, we can now observe solar systems elsewhere in our near galactic neighborhood at various stages of evolution—from dark molecular clouds composed of dust and gas, to new stars, and even to dusty disks around new stars that have indications of planet formation. To this evidence is added the fact that the power of modern computational methods permits us to make *in silico* models of planet formation. In these simulations, thousands of small particles are inserted into a spinning disk around a star, and the computer program keeps track of what happens to the particles according to Newtonian laws of physics. In other words, each particle has a certain mass that exerts a gravitational force on other nearby particles, and each particle also has a velocity as it moves within the slowly spinning disk. In this way, the interactions within the disk can be followed over millions of years of virtual time.

In one such simulation, small aggregates of particles can be seen gathering into larger lumps, which ultimately grow into planetesimals ranging in size from a few kilometers to hundreds of kilometers in diameter. Due to disturbances caused by Jupiter's gravity, the asteroids in orbit between Mars and Jupiter stop accreting at this point, and are considered to be the leftovers of the planet-forming process. After objects the size of planetesimals appear, planet formation can begin in earnest. The vast numbers of planetesimals are in wildly chaotic orbits that frequently produce collisions, and each collision results in a larger body and showers of smaller fragments. In just a few million simulated years, the primary accretion is completed, leaving planets of variable size orbiting the central star. Most of the original planetesimals are swept up by the planets, and any remaining dust is driven into the outer solar system by the light and stellar wind of atomic particles emitted by the newborn star. (See "Sources and Notes" for more information about this simulation.)

WORLDS IN COLLISION: THE EARTH-MOON SYSTEM

The consensus view is that our own solar system originated in this way. However, there is one last event that must be taken into account in relation to the origin of life, which turns out to be the special case of our own planet, Earth. Of all the planets of the solar system, only Earth has an ocean of liquid water, and a moon that is so large in relation to its size. Where did oceans come from? Where did the moon come from? Why is the moon covered in craters produced by the impact of giant asteroid-sized bodies? Was a moon, and the tides it produces, essential for life to begin?

To follow the train of logic that led to the current model, we can first sketch out alternative proposals and show why each is problematic:

1. The moon developed independently as a small planet and was captured by the Earth's gravitational field. However, according to Newtonian laws of gravity and motion, it is mathematically impossible for two moving objects to pair up this way.

2. The moon is just a piece of Earth that broke away due to centrifugal force. Again, this is physically impossible because Earth's rotational velocity is too slow to generate such forces.

3. Earth and its moon congealed as a pair out of the same dust cloud that produced the other planets. The problem here is that Earth has a metallic iron-nickel core, but we can calculate the density of the moon from its known gravity, and it is too light to have an iron core. If Earth and moon formed out of the same material, they should both have metal cores.

4. The orbit of a Mars-sized planet (sometimes called Theia after a Greek goddess) happened to intersect the orbit of proto-Earth. The two planets underwent a cataclysmic collision in which the mineral and metallic contents of Theia were added to that of Earth, bringing it up to its present mass. The energy of the collision melted Earth's crust and sent a small fraction of the total mass into orbit, forming a ring of vaporized rock from which the moon congealed by gravitational accretion.

After much debate and argument, the fourth explanation is now the consensus view. The laws of physics can be used to devise a computational model of the process, and one can watch the formation of a simulated Earth-moon system on a computer screen. The clincher is that the model makes predictions about the mineral composition of the moon. When moon rocks were returned during the Apollo series of lunar explorations, laboratory tests of the rocks' mineral content matched the composition predicted if they were produced by a collision of planets.

What does this mean for the origin of life? The main point is that Earth's crust would be turned into molten lava by the energy released during the collision, and no organic compounds could survive this searing temperature. Most of the early atmosphere, including water vapor, would have been blasted out into space. This means that the atmosphere had to be replaced in some way. Furthermore, all the organic compounds required for life to begin needed to be replaced. But what were the sources of water and organics? And what kinds of organic material would be available?

These questions will be addressed in later chapters, but the short answer is that the minerals composing the interior of Earth still had a lot of water and carbon dioxide left over from the original accretion process, and these were continuously brought to the surface by volcanic outgassing. A smaller amount, now estimated to be about one-tenth of all ocean water, was delivered to Earth by impacting comets, which are giant conglomerates of ice and dust containing 60% to 80% water. The two primary sources of organic material were delivery from a continuing infall of comets, meteorites, and dust, and a synthesis of organic compounds by a variety of geochemical reactions.

CONNECTIONS

As a result of the revelations emerging from today's astronomy and cosmology, our understanding of life is no longer confined to the thin layer of Earth's crust called the biosphere. Instead, we have a much expanded narrative that is encompassed by the new field of astrobiology. For instance, we now know that the primary elements required for life are continuously produced by nucleosynthesis in stars. At the end of their lifetimes, when stars run out of hydrogen that supports nuclear fusion reactions, they first expand into red giants and then collapse into white dwarfs. During collapse they explosively eject much of their mass into the interstellar space, where the atoms aggregate into microscopic dust particles of silicate minerals coated by a mixture of ice, carbon dioxide, and organic compounds. The dust accumulates into dense molecular clouds light-years in diameter, and these clouds undergo gravitational collapse to form new stars, a few of which have a dusty nebular disk from which new planets form. Planets are produced by collisions between kilometer-sized aggregates called planetesimals. Sometimes much larger collisions occur, such as the one that produced our own Earth-moon system. The energy of the collision melted Earth's surface and eroded much of the early atmosphere, which was replaced by volcanic outgassing of water and carbon dioxide from Earth's interior. The molten surface cooled rapidly by radiating heat into space, and the water vapor condensed into oceans.

The continuous infall of meteorites, and their composition, strongly supports the conclusion that the elements required for the origin of life were present in the early atmosphere and oceans, together with immense amounts of energy that could drive the synthesis of ever more complex organic compounds. Chapter 2 will describe the environment of early Earth and address the next question: Where could life begin?

2

WHERE DID LIFE BEGIN?

Our data support recent theories that Earth began a pattern of crust formation,
erosion, and sediment recycling as early in its evolution as 4.35 billion years ago,
which contrasts with the hot, violent environment envisioned for our young planet
by most researchers and opens up the possibility that life got a very early foothold.

BRUCE WATSON, 2005

Our driver turned off the engine of his rumbling Russian Army troop carrier and left us at the edge of a deep canyon carved through layers of ash by a small stream of glacial meltwater. We began to climb, following a faint trail that led toward a peak still hidden in clouds. Vladimir Kopanichenko, our guide, led us up mud-covered slopes, crunching over packed snow and ice among tumbled, house-sized boulders. Despite a chilly wind that whistled past, even in late summer, we were hot and sweaty and often stopped to catch our breath and to take in the amazing scenery around us. When we looked back downhill, ash and lava flows formed hills and valleys, with scattered patches of low shrub in sheltered areas far below. The jagged volcanic landscape of Kamchatka defined the horizon. Above us loomed the blasted peak of Mount Mutnovski, an active volcano that had erupted just a few years earlier.

Two hours later and 2,000 feet higher, we peered over the edge of the crater. It was hard to grasp the vast chaos beneath us. In this lunar landscape of dark gray lava and white snow, there was nothing alive except our team of six scientists. A small glacier covered the other side of the crater. Distant roaring sounds emanated from deep inside as clouds of steam rose into the blue sky.

Earth, air, fire, and water, I thought to myself. These ancient elements brought together here in far eastern Russia are stirred by the heat energy left over from the beginning of our planet's history. Except for the glacier, this place seemed like a remnant from that time, a model of what the entire Earth was like four billion years ago, before life began.

I had come to Kamchatka with two goals. The first was inspired by obscure reports in Russian-language journals that organic compounds, including amino acids, had

FIGURE 3

The author is shown collecting samples from a fumarole in the crater of Mount Mutnovski, an active volcano in Kamchatka, Russia. Such volcanic environments are useful analogues of similar environments on early Earth, where minerals, water, and atmospheric gases meet at interfaces. The energy available at such interfaces can drive chemical reactions that produce more complex organic molecules. Photo courtesy of Tony Hoffman.

been discovered in the boiling springs and vapors of volcanoes in Kamchatka. Everyone agrees that the origin of life required a source of organic compounds, but no one really knows what the primary source might have been. One possibility was that most of the compounds were produced by geochemical synthesis in volcanic regions of early Earth. For instance, laboratory simulations showed that mixtures of carbon monoxide and hydrogen gas, when heated under pressure, could produce hydrocarbon derivatives resembling the compounds that assemble into cell membranes. It would be a real breakthrough if we were able to detect similar reactions occurring in volcanoes today, so I collected samples of volcanic mud to take home for analysis (Figure 3).

The second goal was basically to cover my bet. To finance our trip, we were using a small grant from the NASA Astrobiology Institute—money from American taxpayers. What if we got all the way to Kamchatka and found no organic compounds? That would be embarrassing. For this reason, I had brought along a mixture of organic compounds similar to those we thought might have existed four billion years ago to kick-start the origin of life: four amino acids, a fatty acid, the four bases of nucleic acid, phosphate, and glycerol. We knew that under certain conditions we could get these to react in the laboratory to produce a variety of more complex compounds related to the molecular structure and function of life. I proposed to add these basic compounds to a volcanic pool to see what happens in a natural setting beyond the sterile confines of a laboratory. Most of my colleagues think this kind of experiment is foolish because the conditions are so uncontrolled. But I think of it as a kind of reality check. We can get all kinds of interesting reactions to work in the controlled conditions of a laboratory and then jump to the conclusion that similar reactions must have occurred on prebiotic Earth. But what if we are overlooking something that becomes apparent only when we try to reproduce the reactions in the natural setting we are trying to simulate in the lab?

THE SITE OF LIFE'S BEGINNING

I'll save the results of the experiment for a later chapter, but this mention illustrates how scientists think about the origins of life. We consider all the possible sites that might have been present on early Earth, using today's geology as a guide. We try to simulate those conditions in the laboratory, and then we watch what happens. Because working scientists have only about 40 years of their lives to spend on research, most of us pick a site that we think is plausible and use that assumption to design experiments. But someone from the outside would be astonished by the lack of agreement among experts on plausible sites, which range all the way from vast sheets of ice occasionally melted by giant impacts, to "warm little ponds" first suggested by Charles Darwin, to hydrothermal vents in the deep ocean, and even to a kind of hot, mineral mud deep in Earth's crust. Despite this range of opinion, there are a few things scientists agree about.

The agreements have a special terminology for which we use the word "constraint." Instead of just waving our hands and saying that anything is possible, we agree on a range of environmental conditions that seems highly plausible and then constrain our research within that range. Those conditions are what we try to simulate in the laboratory, and each has a set of physical and chemical laws that characterize it. The goal is to further constrain conditions by experiments in order to find a specific site that produces molecular assemblies having one or more functions associated with life. In my case, I chose to investigate organic compounds that self-assemble into the membranous boundary structures essential for the origin of cellular life. It is something like buying tickets in a giant lottery. We choose from multiple combinations of physical and chemical numbers with the expectation that one set will win the grand prize, which is to produce a system of molecules that has not just one or a few properties of life, but is actually alive. Whoever wins the lottery will have discovered the second origin of life, probably in a laboratory setting, four billion years after life actually began on Earth.

One way to impose constraints is first to agree on where life *cannot* begin, which gives us a basis for considering plausible sites for the origin of life and excludes implausible environments. After all, life has had several billion years to evolve and manages to live in virtually every possible niche that has a certain set of properties. But where are the edges of the biosphere, beyond which life cannot exist?

Let's begin with an easy one: red-hot volcanic lava. With my family, I once visited the lava that is actively flowing down the slopes of Kilouea on the Island of Hawaii. We are used to stones and rocks being "rock solid," but at Kilouea, molten orange-hot rock oozes across a two-lane road and pours over a cliff, causing clouds of steam to erupt from the Pacific Ocean. My daughter Ásta, five years old at the time, was understandably *very* suspicious of the stuff and would not go near the lava flow. It reeked of sulfur and radiated an oven-like heat, even from 50 feet away. Most of our planet is like this lava—even hotter—still molten from radioactive elements disintegrating in the iron-nickel core and the residual heat generated by the violent impacts and collisions that formed Earth 4.57 billion years ago. We live on a thin, rocky crust that is frozen only because it constantly radiates heat into outer space.

Why can't life occur in hot lava? The chemical bonds that hold organic material together begin to break when temperatures rise above a few hundred degrees Celsius, and lava is more than 1,000°C. This fact of chemistry constrains life to temperatures well below that upper bound of chemical stability, to approximately the temperature of boiling water.

At the opposite extreme, can life exist in solid ice? The answer is that life can survive freezing at ice temperatures, but it cannot grow or proceed through a life cycle to produce a new generation of organisms. The reason is simple. Life requires a source of nutrients to grow, which means that the nutrients must diffuse through a liquid medium to reach the cell and be transported across the membrane boundary. Dissolved nutrients must also diffuse within the cell to support the biochemical reactions of metabolism that are essential for life. In ice, free diffusion of nutrients cannot occur, nor can growth or metabolism take place. So a bacterial cell or a sperm sample can survive a deep freeze in liquid nitrogen at 196°C below zero, but none of the usual functions of life can begin until the ice thaws and liquid water is available.

By thinking about the limits of life today, we can constrain the origin of life to a certain temperature range of about 100°C—roughly the range in which liquid water occurs. We can test this constraint by exploring sites where microbial life actually exists today in extreme environments. One such place is the desertlike environment of Antarctic dry valleys, which are both arid and cold. Imre Friedmann, one of the first scientists who might have called himself an astrobiologist, was exploring the dry valleys in the 1980s and discovered that a microbial form of life called cryptoendoliths (a new word appropriately created from Greek words meaning "hidden inside rocks") can eke out a living by inhabiting a thin porous layer beneath the surface of sandstone rocks. These organisms include lichens, algae, and bacteria, which are all photosynthetic, so the microscopic life shows up as a layer of green chlorophyll. But such communities only exist where the temperature occasionally rises above freezing during the Antarctic summer so that a small amount of liquid water appears for a few days every year. Outside of this habitable zone, there is no active life in the arid, frigid deserts of Antarctica.

Going again to the opposite extreme, we might dive in the Alvin submersible to hydrothermal vents that were discovered in 1980. John Baross and his research associates at the University of Washington have carried out such explorations with the aim of finding the highest temperature at which life can survive. Hydrothermal vents occur in the deep ocean, where cracks appear in the tectonic plates that support continents. Sea water circulates down into the hot rock below and dissolves some of the mineral materials. Towering columns are formed when extremely hot water, heated to 300°C and laden with sulfide minerals, seeps up through the fractured rock into the cold sea water. How can water be this hot? Everyone knows that water boils at 100°C, or 212°F. In fact, if 300°C sea water could somehow be brought instantly to the surface, it would explosively vaporize, but deep in the ocean the pressure is so great that boiling cannot occur. What does happen is that the mineral content is much less soluble in cold water, so it precipitates to form a kind of chimney through which the hot water continues to flow. Occasionally, these chimneys grow into towering structures 15 stories high, one of which was dubbed "Godzilla" because of its size and shape.

The chimneys of hydrothermal vents are an ideal place to look for extreme conditions that limit life because the temperature varies from the 4°C sea water outside the vents to the 300°C water on the interior. John Baross and his students broke off one of the mineral columns, transported it back into the lab, and used a microscope to look for bacteria, which they found throughout the porous structure. In earlier studies, they were able to measure the temperature in chimneys and established that bacteria were able to grow even when the temperature was as high as 121°C. This is the record so far, but John believes that it will go even higher.

In 1983, John and his graduate student Sara Hoffman coauthored a paper, which suggested that hydrothermal vent environments might even have been the original site where life began. This idea attracted considerable attention and deserves serious consideration because a variety of interesting chemical reactions can occur under hydrothermal conditions that are relevant to the origin of life.

To summarize, microbial life on Earth can survive and reproduce within the range of a few degrees below 0°C, where salty water can still exist as a liquid, to 121°C in hydrothermal vent environments. The only absolute requirement is liquid water. This range is an example of how constraints are used when we think about the site at which life began on Earth. It is also true that the microbial organisms inhabiting hydrothermal vents are extremophiles. Although a few forms of specialized life can eke out a living in these conditions, living organisms in general do not flourish there. The most abundant and complex forms of life do best in conditions somewhere between these environmental extremes. It may be that even though life can survive in extreme conditions, a more benign environment was required for the origin of life.

TAKING THE MEASURE OF LIFE'S ORIGIN

We can now consider another perspective on the site of life's origin, which is related to the scale that must be taken into account. In general, there are four such scales: global, local, microscopic, and nanoscopic. Let's discuss each of these in turn.

When we "think globally" we are thinking about the general properties of Earth itself: primarily its atmosphere, hydrosphere, and lithosphere. For instance, we know the average temperature of the lower atmosphere with considerable precision because thousands of weather stations around the world have kept track of local temperatures since the early 1900s. When these are averaged over time, including high and low temperatures associated with the seasons, the average global temperature today is 14°C. Furthermore, the average temperature has increased by ~0.8°C in the last 100 years—an increase referred to as global warming. We can also deduce from the geological record that in the past, Earth has been much colder (perhaps even an ice-covered "Snowball Earth" 700 million years ago) or much hotter (during the Carboniferous Era 300 million years ago, when plant life produced the biomass that would later be transformed into massive deposits of coal).

The hydrosphere comprises all liquid water on Earth's surface, which is mostly in the oceans. Sea water is slightly basic, having a typical measured pH of 8.2, and contains about 30 grams of salts in every liter. These are primarily in the form of sodium chloride, or table salt, but a fair amount of calcium and magnesium salts are also present. The pH of the ocean is of great concern just now because even a slight acidification will markedly affect the ability of marine organisms such as coral and mollusks to produce the calcium carbonate mineral of coral reefs and shells. The carbon dioxide produced by human activities in which fossil fuels are burned is a major contributor to acidification because carbonic acid is produced when carbon dioxide dissolves in water. We are certain that liquid water was essential to the origin of life, so pH must also be a significant factor in our scenarios. The pH scale will be discussed later in further detail, but it is easy to remember that pH 7 is the pH of pure water (neutral pH). So pH ranges lower than 7 are acidic, and those above 7 are basic, also referred to as alkaline. In nature, most sea water and fresh water is in the neutral pH range, but in volcanic environments water can be very acidic, close to the pH of battery acid.

The lithosphere is represented by the mineral components of Earth's crust, and today it is composed of those three mineral types that we all learned in high school science classes: igneous (rocks like granite and basalt that are volcanic in origin), metamorphic (rocks like slate and marble that have been transformed by high pressure and temperature), and sedimentary (rocks like limestone and sandstone that were produced from sediments typically in seabeds).

We should also include the concept of the biosphere, or the habitats of all life on Earth, which ranges from mountain peaks 8 kilometers above sea level to the ocean floor, about 8 kilometers below sea level.

What was the global temperature of early Earth's atmosphere? What was the pH of the ocean, and was it salty? What was the lithosphere like at that time? We cannot answer any of these questions with certainty because almost nothing on Earth has survived from four billion years ago due to the geological processing of the crust over time. But observations of neighboring planets and satellites can tell us a lot. For instance, the craters we see on Earth's moon are about four billion years old, and if the moon got pelted with giant impacts, so did Earth. We call that formative period the Hadean Eon, from the Greek word *Hades*. As recently as 10 years ago, the consensus was that Earth was a hellish place from 4.4 to 3.8 billion years ago (Hadean Eon). But recent discoveries, which are described in the next chapter, suggest that an ocean was present much sooner than we thought.

The Atmosphere   In the case of the atmosphere, we can also impose some constraints on our knowledge of the global-scale environment when life began. A clue to the gas composition of the early atmosphere lies in our increased understanding of solar system history. For instance, spacecraft have been placed in orbit around Venus and have even landed there, sending back data for a few minutes before the heat destroyed the instruments. Our sister planet is just a little too close to the sun for comfort. We discovered that the Venusian atmosphere is mostly carbon dioxide, and the amount that is present produces a pressure equivalent to ~90 Earth atmospheres. (To get an idea about what this means, the air pressure in the tires of your car is about 3 Earth atmospheres.) Venus is so hot (460°C, the temperature of molten lead) that there is no liquid water.

Is it possible that Earth also had a carbon dioxide atmosphere four billion years ago? If so, where did it all go? Unlike Venus, Earth has always had oceans in which carbon dioxide could dissolve and precipitate as calcium carbonate, the same mineral that composes limestone, sea shells, and coral reefs. This means that the carbon dioxide of Earth's early atmosphere ended up as rocks, leaving behind mostly nitrogen that remains to this day. Nitrogen gas is composed of two nitrogen atoms linked into a molecule (abbreviated N_2) and is very stable. For this reason, it remained as a gas rather than combining into a mineral as virtually all other elements do.

When we add up all the known carbonate mineral in Earth's crust, it is equivalent to 65 Earth atmospheres, nearly that of Venus. This amount was never present all at once because carbon dioxide was being emitted slowly as volcanic gas over millions of years. The atmosphere at the time life began was mostly nitrogen and carbon dioxide, with trace amounts of carbon monoxide, methane, hydrogen, hydrogen sulfide, and perhaps a little oxygen produced by chemical reactions in localized environments.

The Hydrosphere   The hydrosphere was, and still is, mostly ocean water. As will be described in the next chapter, there is evidence from the study of zircon minerals that oceans were present surprisingly early in Earth's history, perhaps more than four billion years ago. Sea water then had dissolved salt, just as it does today. At this point, we need to review a little elementary chemistry related to concentration. Anything that dissolves is called a solute, and the result is a solution. The standard way for chemists to express concentration is in molarity, which sounds like an invasion of small animals that burrow in our lawns, but it is actually derived from the same word as "molecule." A mole is a unit of measurement equal to 6×10^{23} molecules. When you buy a dozen eggs, you get 12, but if you bought a mole of eggs, you would get about 6×10^{23} eggs, a mass equivalent to that of the moon. (It's astonishing to think that the same number of molecules is present in 3 tablespoons of water, or 18 grams to be exact.)

Concentration is defined as the number of moles of solute per liter of water, and the salt concentration in sea water is a good example. One mole of salt has a mass of 58 grams, so if we dissolve 58 grams of salt in 1 liter of water, the concentration will be 1.0 M, which is pronounced "one molar." For convenience, we sometimes use the unit millimolar (mM): 0.001 M = 1.0 mM. Some solutions are so dilute that we must use units like micromolar (μM), nanomolar (nM), and even picomolar (pM), each 1,000 times less than the preceding concentration.

Getting back to sea water, the salt concentration today is about 0.5 M (500 mM). Were the first oceans also salty? It is possible to add up all the salt that is known to be present in the immense geological formations we call salt mines, which was deposited when early seas evaporated, similar to what is happening to the Dead Sea today. If we could dissolve all of that salt back into the ocean, sea water would be nearly twice as salty as it is today, so it seems likely that the first oceans were also loaded with sodium chloride. There would also be dissolved calcium and magnesium ions (equal to 10 mM and 53 mM respectively today), so sea water would qualify as "hard water," which affects some of our ideas about the chemistry and physics of life's beginnings, as will be discussed later.

Finally, there would be much more dissolved iron in early oceans than there is now. We normally don't think of iron as being soluble, but this is only true of its metallic form. Iron can also exist in a soluble form called ions, as can all other metals, including sodium, potassium, magnesium, and calcium in particular. When a metal like iron is treated chemically such that negative electrons are stripped away from the iron atoms, the resulting ions have one or more positive charges and are very soluble in water. Iron can exist with either two or three positive charges, abbreviated Fe^{2+} and Fe^{3+}. (The Fe comes from *ferrum*, the Latin word for "iron.") The Fe^{2+} form is soluble, but when it comes into contact with oxygen it is oxidized to Fe^{3+} and turns into iron oxide, or rust. The mineral we call iron ore is simply Fe^{3+} that precipitated from the ocean as iron oxide when bacteria capable of photosynthesis finally began to add significant amounts of oxygen to Earth's atmosphere about two billion years ago. If we could dissolve all the iron ore back into the oceans and do the same calculation as we did with salt mines, the iron concentration of sea water would be up in the millimolar range. Today, most of the iron has precipitated as iron oxide, and the rest is avidly taken up by all the life in the sea, which requires iron as an essential trace element. As a result, the iron concentration in sea water is now 31 picomolar, 100 million times less concentrated than the iron content of the early oceans.

Sea water also contains hydrogen ions or protons, by which we gauge the acidity or alkalinity of a solution. This is usually expressed as pH, which is defined as the inverse log of the hydrogen ion concentration. It turns out that pure water is not actually all composed of H_2O, but instead a very tiny fraction is always separated into positive H^+ cations and negative OH^- anions. The concentration of hydrogen ion is 10^{-7} M, where -7 is the logarithm. The inverse of -7 is 7, so the pH of pure water is simply 7, which we call neutral pH. A carbonated soft drink is pH 3, and stomach acid is pH 1—a million times more concentrated in H^+ than water. At the other end of the scale, when you put lye down a plugged drain to clear it out, the pH of the solution is very alkaline, around 14.

What about the pH of sea water? It is slightly alkaline, pH 8.2, but it was probably on the acidic side when life began, between pH 5 and pH 6 in the upper layer of the ocean. This pH range turns out to be pretty important. For instance, at a slightly acidic pH range, organisms are unable to cause biominerals such as calcite (calcium carbonate) and apatite (calcium phosphate) to precipitate into hard structures like shells, bones, and teeth. This might be part of the reason that from 3.5 billion to 1 billion years ago, the biosphere consisted entirely of single-celled bacteria. About 540 million years ago, a point in time we call the Cambrian radiation, something changed that allowed multicellular organisms to proliferate. The Cambrian shows up in the fossil record as vast numbers of multicellular organisms such as corals and mollusks that could produce supporting structures and shells from calcium carbonate.

An important point to make is that all cells today expend considerable energy to regulate their internal environment and typically keep concentrations of certain ions much lower that sea water. If life began in a salty marine environment, how did this regulation come about? Were ionic concentrations important at all for the origin of life? In fact, ionic concentrations strongly affect the most important processes in living cells today. For example, slight imbalances in intracellular pH and calcium ion can be fatal, and imbalances in the concentrations of sodium and potassium across the cell membrane can cause cells to swell and burst. This is why cells expend so much energy regulating the concentrations of these ions. In Chapter 7's discussion of self-assembly processes, I will describe how the presence or absence of certain ions can be highly disruptive, so much so that we must rethink the common assumption that life began in a marine environment.

The Lithosphere   When life began nearly four billion years ago, the lithosphere must have been relatively simple compared to today's world. Not enough time had passed for continents to appear so that most of the land masses rising above the sea were volcanic, like Hawaii and Iceland. This means that the primary mineral components of the lithosphere were basaltic lava and volcanic ash. An important point here is that the appearance of volcanic land masses made it possible for fresh water to exist. Water evaporating from the sea would precipitate as rain and fall on volcanic islands, as it does on Hawaii today. Streams of water would flow down the sides of mountains, carrying with them any organic compounds present in the rain clouds or on the mineral surfaces of the volcanic rocks. These solutes would accumulate in freshwater ponds and be available for further chemical reactions required to begin the life process.

LOCAL SCALES

Although global scales can be constrained within certain limits, local scales have much more latitude in physical and chemical properties. Consider Earth today. The average temperature is 14°C, but local variations range all the way from molten volcanic lava over 1,000°C to Arctic ice 60°C below zero. Even more extreme are the temperatures in air heated by lightning strikes, in which the energy content is so high that many interesting chemical reactions can occur that would not happen under ordinary conditions. I will describe these reactions in Chapter 4 when I discuss sources of organic compounds.

The pH of the ocean today is slightly basic, but localized environments range from very acidic (pH 1) to highly alkaline (pH 11), and salt content ranges from freshwater lakes to evaporating salt ponds in which the concentration is so high that salt crystals precipitate.

What this means for origins of life research is that even though there are constraints on global-scale properties, we can use our imagination at the local scale, and we certainly have. Local environments that have been proposed for the origin of life range all the way from hydrothermal vents in the deep oceans, to evaporating ponds, to glacial ice melted by giant impacts. Every site has advocates, so there is as yet no consensus regarding the most plausible local environment for the origin of life. As will be described in later chapters, the site I am exploring as a plausible prebiotic environment is what we find in the neighborhood of many active volcanoes, such as Mount Mutnovski in Kamchatka. Another such site a little closer to home for me is Mount Lassen, in Lassen National Park, California. This is an area of hot springs, bubbling mud pots, and small geysers called Bumpass Hell, named after Mr. Bumpass, an early explorer who fell through a crust into boiling water and burned his leg badly. I think that geothermal sites like Kamchatka and Bumpass Hell can guide our thinking about what a localized environment on early Earth could have been like. This is the reason I visit such sites, testing whether the simulations we carry out in the laboratory actually work in a complex natural environment.

MICROSCOPIC SCALES

We must also take into account that the first life was microscopic. It may be that we will not understand the origins of life until we better understand possible sites in terms of their microscopic scale. For example, if we examine a chunk of lava, or a sample of the chimney material from a hydrothermal vent, it looks like solid rock. But if we slice such material into very thin strips and examine it with a microscope, we see that it is highly porous, filled with tunnels and caverns ranging from the size of bacteria to small holes visible to the eye. NASA researcher Michael Russell has suggested that such mineral compartments could have served the same purpose as membranes in confining the reactions of early life to a limited space. Other examples of microscopically porous minerals include clays and a variety of sedimentary deposits that would be expected to occur wherever ash fell into standing water and intertidal zones.

Cellular life is microscopic, with measurements of living cells ranging from bacteria a few micrometers in diameter to amoebae several hundred micrometers across, but the actual molecular components of life are measured in nanometers (nm). Only 0.2 nanometers in diameter, water is the smallest molecule involved in life processes. Glucose is about 1 nanometer in diameter, and a protein like hemoglobin of blood is 4 nanometers in diameter. For life to begin, there must have been processes occurring at the nanometer scale. Probably the most important would be thin films of organic material adsorbed to and organized by interfaces, which are defined as the surface between a solid and a liquid, or a liquid and a gas. As will be discussed later, one hypothesis is that life began when chemical reactions related to metabolism were initiated on the nanoscopic interface of an iron sulfide mineral called pyrite. Another hypothesis is that organic compounds could adhere to clay mineral surfaces in such a way that they produced orderly arrays containing a primitive form of genetic information. Later in this book, I will describe how nanoscopic structures formed by lipid molecules can add order in this way and promote synthesis of long molecules resembling nucleic acids such as RNA and DNA.

THE CHANGING EARLY EARTH ENVIRONMENT

There are two final points to make about our understanding of the site of life's origins. We typically discuss this topic as a kind of snapshot of plausible sites, but in reality, global conditions dramatically changed over the first billion years of Earth's history, and these changes, in turn, affected the physical and chemical conditions of the localized sites where life began. Some of those changes were accompanied by immense amounts of energy released into the environment, and we must take into account that whenever energy is available, it drives chemical reactions that cannot occur in environments that are near equilibrium. Other chemical reactions driven by energy fluxes can destroy complex chemical compounds required by living organisms. For instance, when red-hot lava flows down the side of Kilohuea and buries plant life, the chemical bonds linking the polymers of life are broken by the heat energy and are dispersed as gases so that chemical complexity becomes chemical simplicity. Going in the other direction, if a mixture of dry amino acids is heated to less extreme temperatures, say 90°C, the amino acids lose water over time, and they link up into long polymers resembling proteins. Such reactions are called condensation, and the overall process is classified as synthesis because new chemical bonds are formed. In this case, the heat energy and dry conditions cause complexity to emerge from a simple mixture of amino acids. The most important point is that on early Earth, just as today, synthetic reactions and degradative reactions occurred simultaneously. In order to understand the origins of life, we need to find conditions that can drive chemical reactions energetically uphill, toward complexity, keeping in mind that the rate of synthesis must be fast enough for the products to accumulate before they undergo degradative reactions and return to monomers.

Now we can outline the kinds of global-scale changes that would occur during the half billion years between the initial formation of Earth and the origin of life. Enormous amounts of heat energy were generated by the accretion process, including the moon-forming event. When would the surface become cool enough for water to condense out of the atmosphere as rain? The surprising answer was found by researchers in Australia, the University of Wisconsin, and the Rensselaer Polytechnic Institute in New York, who studied the composition of zircons. These diamond-like minerals are very tough so that after they form, they are stable for billions of years. Furthermore, their chemical composition provides a kind of thermometer. For example, a zircon is composed of zirconium, silicon, and oxygen ($ZiSiO_2$), but it can have variable amounts of titanium (Ti) as well. Bruce Watson, working at Rensselaer, found that the titanium content of zircon is directly related to the temperature at which the mineral forms. Until recently it was thought that the surface of Hadean Earth was basically a sea of molten rock, but the titanium content of zircons that crystallized in the Hadean Era is consistent with a temperature of 688°C, characteristic of a hydrated zircon structure rather than a zircon formed in lavalike temperatures of 2,000°C.

Let's assume that the conclusion is correct so that liquid water was available very soon after the moon-forming event, which occurred about 4.4 billion years ago. Could life have sprung up immediately? The answer is that it could have, but there were still some major hurdles to overcome, which are associated with what we refer to as the Late Heavy Bombardment that will be described in the next chapter.

AN ICY ORIGIN?

There is an alternative to a hot site for the origin of life. Jeffrey Bada, collaborating with Stanley Miller at the University of California, San Diego, took into account the fact that amino acids and other soluble compounds required for life have finite life spans in water solutions. Furthermore, the half-time of degradation is strongly related to the temperature, as might be expected from common experience. It can be estimated that the entire ocean volume passes through high-temperature hydrothermal vents every 10 million years, and the high temperatures associated with the vents (300°C and higher) would cause amino acids to break down into smaller fragments of little use for the origin of life. However, if the same amino acids were frozen at ice temperatures, they would last for much longer times, measured in thousands to millions of years. The same is true for nucleotides, the monomers of nucleic acids. Hydrocarbons are much more stable over time. This is why we still can discover oil and other fossil fuels that were deposited 300 million years ago from the remains of abundant plant life buried as sediment and then processed by the high temperature and pressure associated with sedimentary mineral formations thousands of feet beneath Earth's surface.

But how could ice have been present on a hot early Earth? There is a consensus that Earth was initially hot, and that oceans rapidly formed as it cooled down, but a relatively warm global temperature at the time of life's origin is still an assumption. There is also a consensus that, at that time, the sun only emitted 70% of the energy that it does today. This is called the "dim early sun paradox" because it is not hard to calculate that the oceans should have frozen solid! To resolve this dilemma, astronomer and author Carl Sagan proposed that the abundant carbon dioxide in the early atmosphere might have acted as a greenhouse gas to keep Earth warm enough so that the oceans did not freeze, but this remains a conjecture. In fact, 750 to 600 million years ago, just before the Cambrian radiation began, our planet might have passed through a period referred to as "Snowball Earth," when ice sheets reached all the way to tropical latitudes. The evidence for the "Snowball Earth" period is geological deposits on every continent that are consistent with extensive glaciation. Computer simulations based on the geological record suggest that extensive melting and thawing occurred as many as four times during this period.

So, if global ice sheets were present as recently as 600 million years ago, why not four billion years ago, when the sun was delivering significantly less heat energy to Earth's surface? This is a reasonable speculation, but what does it have to do with the origins of life? For one thing, in a paper published in 1994, Jeffrey Bada and Stanley Miller argued that organic compounds would not be degraded by heat energy if early Earth was cold enough to freeze over. Equally important is that when sea water containing organic solutes freezes, it does not simply turn into a solid block of ice. Instead, the growth of ice crystals excludes dissolved compounds from the crystal structure, and those compounds become concentrated in thin layers surrounding the crystals. The layers can remain fluid even at temperatures well below the freezing point of water. Not only can ice preserve organic compounds, but the concentration of solutes can promote synthetic chemical reactions that cannot occur in dilute solutions.

CONNECTIONS

Today Earth's atmosphere is composed of approximately 78% nitrogen, 21% oxygen, and 1% argon; but four billion years ago, there was little if any free oxygen. There was much more carbon dioxide present, which slowly declined as it dissolved in the ocean and precipitated as calcium carbonate mineral. The oceans were salty, with more dissolved iron than today's oceans, and volcanic land masses rose above sea level. The average global temperature was in the range of 60°C to 80°C.

A broader range of conditions would be possible at the local scale. Land masses rising above sea level would collect precipitation in the form of rain, producing freshwater streams, ponds, and even lakes. This fresh water would not be permanent. Instead, it would undergo drying and wetting cycles, particularly in the neighborhood of volcanic geothermal areas. Other common local scales would be extensive regions of marine hydrothermal vents and intertidal zones resembling the black sand beaches of Hawaii today. At some elevations and latitudes, there might even have been accumulations of ice, but the extent is uncertain.

The dominant microscopic scale would consist of porous minerals in the form of lava and volcanic ash. Less common would be iron sulfide minerals associated with hydrothermal vents. There would also be thin layers of sand or clay minerals, exposed to the atmosphere. The pores and surfaces of the lava, sand, and ash would be able to act as microscopic test tubes, where a variety of chemical reactions could occur that would not be possible in open water. A second important microscopic-scale condition is that some of the organic compounds present would be relatively insoluble in water. These include oily chemicals with hydrocarbon chains, which would form spherical structures resembling oil droplets.

The most important nanoscopic environments would be films of highly concentrated organic compounds. These would be present at mineral-water interfaces and air-water interfaces in the form of thin films a few nanometers thick. Some of the films would be produced by evaporation or by a process called adsorption, in which the molecules adhere to mineral surfaces. In either case, chemical reactions could occur in the films that could not take place in dilute solutions. Other soaplike molecules could self-assemble into nanoscopic micelles or microscopic compartments resembling cell membranes.

The bottom line is that we cannot be certain of the global and local environment that was the site of life's origin. All we can do is to try to simulate a variety of plausible prebiotic conditions in the laboratory and see which is most conducive for the self-assembly processes that must have led up to the origin of life. We can also study modern examples of such sites in the field, from volcanoes to glaciers, testing the ideas developed in the laboratory. The history of science suggests that a continuous, focused effort to try to understand a problem, even as complex as life's beginning, leads toward increasingly complete knowledge of the factors involved and ultimately produces a satisfactory explanation. We will never know exactly how life began on early Earth, but we will know how life can begin on a suitable planetary surface, because we will watch life emerge when just the right set of conditions come together. The rest of this book describes progress toward this goal.

3

WHEN DID LIFE BEGIN?

Although we are therefore mightily constrained in our efforts to document, and decipher, the Precambrian history of life, we can take one more step toward understanding—we can make an educated guess. This is not as wild and woolly an exercise as it might first appear because a really good guess might cause us to ask new questions that lead to increased knowledge.

WILLIAM SCHOPF, 1991

Much of what we know about the origins of life is based on educated guesses, but a few things are reasonably certain. For instance, with some confidence I can state that the solar system is 4.57 billion years old, and Earth is about 4.53 billion years old. Easy to say, but how can we know for sure? And why isn't Earth as old as the solar system? The reason we have a certain amount of confidence in such numbers is that over the past century, physicists discovered radioactivity and began to understand how radioactive elements could change into other elements over time. We take this knowledge for granted now, but only in the last 50 years have we discovered how to use natural radioactivity as a precise time-keeping method.

A time interval of four billion years is almost incomprehensible. However, we can get a feeling for what this really means by imagining that time passes at 1,000 years per second. At that rate, all of recorded history would zip by in 5 seconds, and prehistoric cave drawings in France were being made 30 seconds ago, when *Homo sapiens* and *Homo neanderthalensis* were still competing for territory. (The Neandertals lost.) Extensive glaciation (ice ages) occurred every 100 seconds, with warmer interglacial periods lasting 30 seconds. (The last ice age ended 11 seconds ago, so enjoy the interglacial warmth while you can.) The first human beings appeared in Africa about 3 minutes ago (200,000 years), and the first human-like primate ancestors about an hour ago (4 million years). Then there is a big jump in time back to the giant impact of a small asteroid that killed off the dinosaurs 18 hours ago (65 million years). Dinosaurs first appeared 200 million years ago and ruled the biosphere for 37 hours of our accelerated time frame. Compare this to the 3 minutes that the humans and their ancestors have been around! But after the giant impact, small, rodent-like mammals had a chance to try their luck, and human beings are the result of that opportunity.

For the next time interval, we need to use days instead of seconds, minutes, and hours. About 7 days back in time we reach the Cambrian Era, 543 million years ago, with the appearance of marine organisms large enough to leave recognizable fossils. Now we can appreciate the largest time span of all, called the Precambrian Era, which goes back to the origin of life. Again in our accelerated time frame, this occurred 43 days ago, equivalent to 3.8 billion years.

For me, at least, this gives a sense of the immense time interval between the origin of life and today's biosphere: 43 days for life to begin and evolve to fill every niche in the biosphere, with just 5 seconds of human history tacked on at the end. Why did it take so long (3 billion years) for single-celled life to evolve into larger multicellular organisms? And after all that time, how could multicellular plants and animals appear so quickly in the evolutionary time scale, about half a billion years ago? There are several reasons, but the availability of oxygen was probably a major factor. Bacteria remained as single cells when light and low-energy chemical reactions were the only energy sources. But photosynthesis by cyanobacteria began to produce oxygen soon after life originated, and little by little, the oxygen accumulated in the atmosphere. When oxygen reached a certain level, about two billion years ago, another kind of bacteria learned to use high-energy oxidative metabolism as an energy source. This fundamental split between cells that use light energy and cells that use oxygen marked the turning point in evolution that eventually led to plants and animals.

A GEOLOGICAL TIME SCALE

Scientists don't think about time scales in terms of 1,000 years per second. Instead, we give names to specific time intervals, and after you get them memorized, they become as familiar as a map of your home town. Some of the names refer to specific points in the geological record, and others to specific points in the biological record. For our purposes there are only a few that need to be kept in mind because these are associated with the origin and early evolution of life. The longest time intervals, or eons, are called the Hadean, Archaean, and Proterozoic. Chapter 2 described evidence from ancient zircons that liquid water was present as oceans more than four billion years ago, in the Hadean Eon, and the first faint traces of life appear about 3.8 billion years ago, in the Archaean Eon. Microfossils of bacterial life have been reported in ancient rocks that are 3.5 billion years old, although there is a certain amount of controversy because nonbiological explanations have also been proposed to explain the structures. Unmistakable microfossils of bacteria become abundant in the Proterozoic Eon, which began 2.5 billion years ago.

To get back to the original question, how can we determine the age of something that is several billion years old? One of the most important methods depends on a remarkable relationship between lead and uranium. There is an interesting history related to these two metallic elements. Uranium the metal and Uranus the planet are both named after the Greek god Uranus, who in mythology ruled the sky. Uranus was the father of Saturn and grandfather of Jupiter, names chosen very appropriately for two other major planets. Uranus was discovered by William Herschel in 1781, and eight years later, Heinrich Klaproth isolated uranium the element, naming it in honor of the new planet. The abbreviation U is obvious, but what about lead and its abbreviation Pb? The metal lead was discovered thousands of years ago, and the word "lead" comes from an early form of the English language. The Latin word for lead is *plumbum*, from which the word "plumber" and the abbreviation Pb are derived. (Lead is a soft metal that is easily shaped and molded, so it was used in Roman times to make pipes for plumbing.)

There is more to the story, which shows how major advances in science are often revealed by a chance observation. In 1896, the French scientist Antoine Henri Becquerel was trying to figure out why certain minerals were phosphorescent (glowing in the dark after exposure to bright light). He thought it might have something to do with X-rays, which had recently been discovered. X-rays were able to penetrate opaque materials and produce an image on film, so Becquerel wrapped photographic plates in black paper and placed different phosphorescent minerals on them for several days. When he later developed the plates, all but one of the plates were completely blank. The exception was the plate exposed to a mineral containing uranium, and Becquerel was astonished to find that an unknown form of energy had penetrated the black paper and produced a smudgy image where the uranium ore had been placed.

This is an example of how a single scientific observation can change the course of human history, both for better and for worse. The penetrating energy attracted the interest of other scientists, including a young Polish woman named Marie Sklodowska, who was fascinated by physics. It was impossible at the time for a woman to get an advanced degree in Poland, so she enrolled at the Sorbonne University in Paris. Marie had heard about Becquerel's results, and she decided to isolate the mysterious component that produced the effect. During the course of her studies, one of Marie's professors, Pierre Curie, noticed her deep intellect and caring personality. Marie, in turn, was impressed by his passion for science. As is said today, they were soul mates. After only a few meetings, Pierre wrote to Marie asking her to marry him:

> It would, nevertheless, be a beautiful thing in which I hardly dare believe, to pass through life together hypnotized in our dreams: your dream for your country; our dream for humanity; our dream for science. Of all these dreams, I believe the last, alone, is legitimate. I mean to say by this that we are powerless to change the social order. Even if this

were not true we should not know what to do. . . . From the point of view of science, on the contrary, we can pretend to accomplish something. The territory here is more solid and obvious, and however small it is, it is truly in our possession.

So Marie Sklodowska became Marie Curie, and together the young couple went on to isolate the radioactive elements polonium and radium. Marie, in fact, invented the term radioactivity to describe their unique properties. To honor their discoveries, in 1903, the Curies were the first husband-and-wife team to be awarded Nobel Prizes.

The Curies isolated tiny amounts of polonium and radium from many kilograms of an ore called pitchblende, but the major radioactive component of pitchblende is uranium. Many tons of uranium have now been produced by extraction from pitch-plende, and the metal has a number of uses other than in nuclear reactors for power generation. For instance, on a shelf in my lab sits a small glass bottle filled with a yellow powder, uranium acetate, that we use to stain biological materials to be examined by electron microscopy. The uranium acetate is mostly composed of the uranium isotope U_{238}, which is present virtually everywhere in Earth's mineral crust. But if I take off the lid and hold a radiation counter above the powder, the buzz from the speaker tells me that the powder is weakly radioactive. Uranium metal is the heaviest naturally occurring element, with 92 protons in its nucleus, and as many electrons in orbit around the nucleus. The technical abbreviation is $^{92}U_{238}$. The 92 is the atomic number (92 protons in the uranium nucleus) and 238 is the atomic weight (92 protons added to 146 neutrons in the nucleus). U_{238} composes 99.3% of natural uranium, while U_{235} (143 neutrons) is about 0.7%.

The main properties of elements are associated with their atomic number, but like uranium, most elements can have different atomic weights due to varying numbers of neutrons. The atomic weight defines an isotope of the element, and often one of the isotopes is radioactive. For instance, there are two natural isotopes of hydrogen in water (H_2O). The first is ordinary hydrogen with a single proton and no neutrons in its nucleus, and the second is called deuterium, which has both a proton and a neutron in its nucleus. There is also a radioactive isotope of hydrogen called tritium, which has two neutrons in its nucleus, but this is made artificially in a nuclear reactor. Another example is carbon, in which the most common isotope is C_{12}, followed by C_{13}, with one extra neutron. There is also a radioactive isotope of carbon with two extra neutrons—C_{14}—which can be used to determine the age of organic, carbon-containing material.

Uranium is surprisingly abundant in Earth's crust—in the same range as tin, mercury, and arsenic. It is common knowledge that if enough U_{235} can be gathered together in one place (about 7 kilograms, the size of a tennis ball), a chain reaction occurs that releases a fraction of the stored energy according to Einstein's famous equation: $E = mc^2$. If the energy is released slowly under controlled conditions, it is available as useful heat in a nuclear reactor. If it is released in a few milliseconds, it produces the massive explosion of an atomic bomb.

The isotope U_{238} cannot sustain a chain reaction, but still undergoes radioactive decay at a certain, very slow rate. If a sample of pure U_{238} metal could somehow be analyzed 4.47 billion years later, precisely half of the U_{238} would have turned into an isotope of lead (Pb_{206}). U_{235} undergoes radioactive decay about six times faster so that a sample of pure U_{235} would only take 704 million years for half of it to turn into a second lead isotope (Pb_{207}) by radioactive decay. This time interval is called the half-life, and is one of the most important ways to establish the age of the solar system and Earth.

THE URANIUM-LEAD CLOCK

How do we determine ancient age by measuring uranium and lead composition? The actual technique is moderately complicated because it involves a method called mass spectrometry. This method allows us to separate all of the isotopes of uranium and lead that are present, plug the results into an equation, then solve the equation to get the age of the sample. But to put it simply, if I knew that a mineral started out long ago with pure U_{238} in it and then determined that the amount of lead in the mineral equaled the amount of uranium, I could conclude that the mineral was 4.47 billion years old. Small amounts of lead are everywhere, so we need to start off with something that contains only pure uranium. It was discovered that zircon, the same mineral that led us to determine that oceans were present over four billion years ago, can incorporate uranium atoms, but not lead atoms into its crystal structure. This means that if we isolate a zircon from a mixture that we want to date, we can assume that only uranium was present when the zircon formed, and that lead atoms would be the result of radioactive decay of U_{238} and U_{235}.

What samples should we analyze with this powerful technique? Why not try to find the oldest rocks on Earth? Earth has undergone extensive geological changes since its birth as a planet, as new continental crust emerged from the underlying mantle and floated on the relatively fluid rock beneath. The continents today are relatively recent features. If you could view a map of the continents only 200 million years ago, when dinosaurs ruled Earth, most of the land mass would comprise a single supercontinent called Pangaea. Over the ensuing millennia, Pangaea separated into a series of plates that support continents today. We also know that new ocean floor forms by an upwelling of magma through a series of cracks in the sea floor that then dives back down by subduction at the edges of continents. The cracks produce the hydrothermal vents described in the previous chapter, and the subduction process along the edges of continents also permits molten magma to reach the surface to produce volcanoes. The Pacific "Ring of Fire," which runs along the coasts of North and South America, extends through Alaska, and then stretches down through Kamchatka, Japan, and Indonesia, is the best known example of volcanism due to subduction.

As a result of tectonic turnover, most of Earth's original crust has disappeared in the past four billion years. Fortunately for geologists, tiny remnants remain in Greenland and northern Canada, in parts of South Africa, and in Western Australia. In 2003, John Valley, a geologist at the University of Wisconsin, traveled to Australia with a team of his students and took samples from a group of rocks that had already been dated by the uranium-lead method at 3.4 billion years old. The rocks, like many such geological formations, were produced from a conglomerate of mineral components that had been mixed by erosion into a thick layer. John's team toted many kilograms of sample material back to their laboratory and laboriously ground the samples into a coarse powder. Then they sorted through the material with a microscope, looking for rare zircons the size of pinheads.

The next step was to examine each zircon with an instrument called an ion microprobe. In this device, a beam of high-energy ions, such as cesium, is directed onto the surface of the mineral so that the component atoms are knocked off into the vacuum. The atoms gain an electrical charge during this process and are called secondary ions. At first, they are a mixture of all the atoms in the sample, including trace amounts of uranium and lead, as well as oxygen isotopes having atomic weights of 16, 17, and 18. The ions are captured by a strong electrical voltage and sent down a column where they pass through a magnetic field that separates them according to their atomic mass. Heavy atoms like uranium and lead are only slightly affected by the magnetic field, but because uranium isotopes (U_{238} and U_{235}) are heavier than lead (Pb_{206} and Pb_{207}), all four isotopes are separated and appear as four peaks in the readout from the instrument. The oxygen isotopes are much lighter and more strongly affected by the magnetic field, but they also produce three distinct peaks. The relative heights of the uranium and lead peaks are used to calculate the age of the zircon, and the oxygen peaks are used to deduce the temperature at which the zircon was formed.

John Valley and his students were astonished when the results of their calculations first popped up on their computer screens. The zircon was 4.4 billion years old, the oldest mineral yet discovered, nearly as old as Earth itself. Furthermore, the oxygen isotope composition of the zircon indicated that it had not formed at volcanic temperatures, but instead at much lower temperatures consistent with the presence of water.

METEORITES AND THE AGE OF THE SOLAR SYSTEM

In a sealed glass container in my lab, I keep a collection of perhaps a dozen meteorite samples. People who do my kind of science know many meteorite names by heart: Murchison, Murray, Allende, Orguiel, Ivuna, and Tagish Lake. These are famous because they are carbonaceous chondrites, a name derived from the fact that they contain organic carbon compounds mixed in with microscopic particles of silicate minerals. Also scattered throughout the black mineral matrix of these meteorites are white pebbles called chondrules. These were produced when the sun first began to produce heat and light. The heat melted aggregates of dust particles, which then fused into chondrules.

Carbonaceous chondrites, along with comets, are samples of the most primitive components present in the early solar system. Other meteorites in my collection are protected in small, stainless steel containers with tightly fitting lids. These are pieces of meteorites found in Antarctica, also carbonaceous chondrites. There is one special container in the collection, and when we have visitors to the lab I shake the container to make the contents rattle. "Hear that?" I ask. "That's a piece of the planet Mars." And it really is. It is a 1-gram sample of the Nahkla meteorite that fell in Egypt in 1906, which was carefully divided into portions and made available to researchers. There are only a few meteorites known to originate on Mars, called SNC meteorites from the first initials of the places where the first three were found. (The N stands for Nahkla, for instance.) We know they are not ordinary meteorites because their mineral composition is what one would expect of a planetary surface, rather than the asteroids where most meteorites originate. The most convincing evidence that the SNC meteorites are from Mars is that they still contain gas, and when this gas is analyzed, it closely matches the atmospheric gas composition of Mars.

The oldest zircons found in terrestrial rocks give evidence that they were produced 4.4 billion years ago, yet we said that the actual age of Earth is 4.53 billion years, and the solar system assembled from a nebular disk 4.57 billion years ago. Where do those numbers come from? The age of the solar system is determined by assuming that asteroids are leftovers from the original accretion process. As described in Chapter 1, meteorites are chunks of asteroids that have constant traffic collisions with other asteroids so that fragments of their surfaces are knocked off into space. A tiny fraction of the pieces happens to intersect with Earth's orbit around the sun. The meteors, or "shooting stars," we see at night are produced by particles the size of sand grains, mostly the dusty remains that comets shed as they pass through the inner solar system. The word "meteor" comes from a Greek word describing aerial phenomena, so meteorology is the study of weather, and a meteor is simply something that happens in the sky. Shooting stars are common, sometimes arriving as showers like the Leonids or Perseids that occur regularly every year. More rarely, a chunk of material ranging in size from a golf ball to a small boulder enters the atmosphere and becomes a fireball as it burns up in the atmosphere in a few seconds. Fragments of the original object that survive the fireball and reach the ground are called meteorites.

About once a century, a much larger object enters the atmosphere and releases energy equivalent to the explosion of a nuclear weapon. One such event occurred in 1908 over Tunguska, Siberia. There is no crater, so the Tunguska explosion was produced by a large chunk of relatively fragile material similar to the Murchison meteorite that entered the atmosphere and exploded before it reached the ground. The explosion over Tunguska produced a massive shock wave, a pulse of sonic energy that traveled downward and leveled 1,000 square miles of forest. Here is an account recorded in 1926 by one of the first scientists to visit the remote area, who interviewed an eyewitness living about 40 miles south of the impact site:

At breakfast time I was sitting by the house at Vanavara trading post facing North. I suddenly saw that directly to the North, over Onkoul's Tunguska road, the sky split in two and fire appeared high and wide over the forest. The split in the sky grew larger, and the entire Northern side was covered with fire. At that moment I became so hot that I couldn't bear it, as if my shirt was on fire; from the northern side, where the fire was, came strong heat. I wanted to tear off my shirt and throw it down, but then the sky shut closed, and a strong thump sounded, and I was thrown a few yards. I lost my senses for a moment, but then my wife ran out and led me to the house. After that such noise came, as if rocks were falling or cannons were firing, the earth shook, and when I was on the ground, I pressed my head down, fearing rocks would smash it. When the sky opened up, hot wind raced between the houses, like from cannons, which left traces in the ground like pathways, and it damaged some crops. Later we saw that many windows were shattered, and in the barn a part of the iron lock snapped.

Every 100,000 years or so, something even larger plows through Earth's atmosphere in a few seconds and delivers enough energy to produce a gaping crater resembling those we see on the lunar landscape. The most recent of these crashed into the Arizona desert about 50,000 years ago. Traveling at 13 kilometers per second, it produced a crater more than 1 kilometer in diameter and 150 meters deep. The energy of the explosion is estimated to be 150 times greater than that of the atomic bomb tested in July 1945 at the Trinity Site near Alamogordo, New Mexico. When the crater was first discovered in Arizona, no one knew what had caused this weird geological feature, so it was named the Canyon Diablo Crater after a small nearby canyon, and it was generally believed to be the result of some kind of volcanic eruption. But in 1903, a mining engineer named Daniel Barringer looked at all the fragments of iron scattered around the crater and correctly concluded that it was the result of a meteorite impact. In scientific literature, it is now called the Barringer Crater in honor of the engineer's good guess, and when you drive by the site on the way to nearby Flagstaff, you will see a road called Meteor Crater that leads to the privately owned tourist attraction.

In the early 1950s, Clair (Pat) Patterson, a Cal Tech geochemist, realized that the particular composition of the iron and lead minerals in Barringer Crater provided a way to estimate its age with remarkable accuracy. The age was determined from uranium-lead isotopes, and in 1956, Patterson published a paper in which he convincingly showed that the fragments were 4.55 billion years old. Since that time, hundreds of estimated ages of other meteorites have been determined, with ages ranging from 4.53 to 4.58 billion years. This range does not arise from experimental error, but instead reflects the fact that the parent bodies of the meteorites must have formed at slightly different times in the early solar system's history. To summarize, the oldest rocks on Earth are zircons, and their uranium-to-lead age is 4.53 billion years. The current scientific consensus is that the age of the Barringer meteorite represents a reasonable estimate of the age of the solar system, slightly older than Earth at 4.57 billion years. Given those numbers, we have a context for attempting to give a date to the origin of life, but first we need to describe how giant impacts in the Hadean Eon were likely to have affected the actual time that life could begin.

LIFE AND THE LATE HEAVY BOMBARDMENT

When a few kilograms of lunar rocks were returned by the Apollo astronauts in the 1970s, geologists were delighted to have actual samples of the moon to study. They found that virtually all of the rocks could be classified as igneous, meaning that they had formed at temperatures of molten magma, but they also found that some of the rocks were obviously pieces of the lunar crust produced by giant impacts of asteroid-sized objects. Anyone with a pair of binoculars can still see the enormous lunar craters that resulted from these impacts. But a puzzle soon arose. When the rocks were dated, they had a range of ages all the way back to 4.5 billion years old, as expected. However, an unexpected abundance of rocks were found with ages around four billion years ago. It was as though something stirred up the solar system during that time interval and sent millions of objects ranging up to hundreds of kilometers in diameter falling toward the sun, where they collided with the inner planets. Some of the craters we see on Mars, Venus, Mercury, and our moon are a permanent record of the bombardment.

Earth could not escape, and the only reason its surface lacks a similar number of craters is that the combination of erosion, plate tectonics, and geological processing has almost completely resurfaced the planet. Nonetheless, there is now a consensus that another bombardment of comets and asteroid-sized objects pelted Earth between 4.1 and 3.8 billion years ago. This period is called the Late Heavy Bombardment (often abbreviated LHB). By extrapolation from the lunar cratering record, it has been estimated that during the LHB, impactors the size of the dinosaur-killer asteroid hit early Earth every 100 years or so. In the same time interval, there were about 40 giant impacts that would have left craters around 1,000 kilometers in diameter, the size of the largest craters on the moon. There would also have been a few impactors that delivered enough energy to produce ocean-sized craters, vaporizing whatever oceans existed at the time and sterilizing Earth's surface. Kevin Maher and David Stevenson at Cal Tech were first to point this out and coined the phrase "impact frustration of the origin of life." In other words, life could have originated on multiple occasions, but only after the last major impact event would it be able to survive to begin the long evolutionary pathway to life today.

If this scenario is correct, we can predict that the first evidence of life will not be older than 3.8 billion years, just about the time the LHB ended. But obvious fossils did not appear in rocks until the Cambrian explosion three billion years later. What evidence could there possibly be for the existence of life that long ago? This question brings us to the concept of biomarkers and biosignatures, which is also central to our search for possible life on Mars or other habitable planets.

FOSSILS, BIOSIGNATURES, AND BIOMARKERS

Most people are familiar with one kind of biosignature called fossils, which are impressions or mineralized structures in sedimentary rocks that are produced by the hard shells and bones of early animal and plant life. Such fossils can be found in Cambrian rocks half a billion years old, then suddenly disappear in older pre-Cambrian rocks. But there is another kind of biosignature which only became apparent when instruments called gas chromatographs and mass spectrometers became available. When living organisms go through the chemical reactions needed to support life, they leave behind traces of their chemistry. There is a good reason why coal and oil are called fossil fuels. During the Carboniferous Period 300 million years ago, the climate was very warm, and immense masses of plant life thrived in the swampy, tropical jungles of the world. This plant life formed deep layers of peat-like deposits, which over time, were buried by layers of mineral sediments. Heat and pressure progressively altered the peat, finally producing the hard energy-rich substance we call coal. Fossil petroleum has a different, less well-defined history. The oil we pump out of the ground was originally produced from the remains of abundant marine microorganisms that fell to sea floors as sediment. Like all life, the phytoplankton cells stored energy in droplets called fat, and the droplets were deposited along with the cells as layers of sediment. Again under heat and pressure, the organic fat was "cracked" and hydrogenated, producing the long-chain hydrocarbons we call oil. To give some idea of the scale of oil deposits, every gallon of gasoline we use today required approximately 100 tons of phytoplankton to turn into fossil fuel.

The reason for explaining all this is to make the point that the most convincing evidence for a biological source of petroleum is that oil contains a set of hydrocarbons that can only have been synthesized by living organisms. These are called terpenes, hopanes, and steranes—the tough remnants of molecules that were once part of cell membranes.

As paleobiologists began to realize in the 1960s that it might be possible to find actual microscopic fossils of the earliest life, they began to search for remnants of Archaean Earth, which could preserve such biosignatures. By chance, certain rock formations in Western Australia, South Africa, Greenland, and Canada escaped the extensive tectonic processing that altered most other rocks at Earth's surface. Geologists had already found several such areas, giving them exotic names that are now famous, at least among origins of life researchers: Isua (Greenland), Gunflint (Canada), Fig Tree (South Africa), Spitzbergen (an island off the Norwegian coast) and Pilbara, Warawoona, and North Pole (in Western Australia). The oldest Canadian and Greenland rocks are 3.8 to 4 billion years old, but have undergone so much diagenesis (defined as geological processing by heat and pressure) that there is no hope of finding preserved fossils. However, Australian rocks that are 3.5 billion years old appear to have fossil stromatolites, layered mineral structures that today are known to be produced by bacteria growing in colonies called mats. These Australian rocks looked like a very promising place to search for the oldest microfossils, particularly because Stanley Awramik, Elso Barghoorn, and Andrew Knoll had already discovered clear evidence of microfossils in the 2.2-billion-year-old Gunflint chert of Canada.

Elso Barghoorn at Harvard pioneered studies of Precambrian microfossils. As early as 1966, Barghoorn and his student William Schopf had published a paper in *Science* that used organic biomarkers in an investigation of apparent microfossils they discovered in South African black cherts dated to be 3.1 billion years old. Barghoorn and Schopf also found pristane and phytane in the same chert, which supported the claim that the microscopic rodlike forms they observed were the remains of early bacteria, rather than inorganic structures. In more recent work, Roger Summons at MIT and Roger Buick at the University of Washington made extensive studies of the fossils and traces of organic compounds that they extracted from Western Australian rocks that are 2.7 billion years old. In a paper published in 2002, Summons and Buick reported that hopanes and steranes could be detected in those rocks, which suggested that photosynthetic microorganisms called cyanobacteria were producing oxygen more than two billion years ago. More surprising is the presence of sterane, a breakdown product of cholesterol that is present only in eukaryotic cells having nuclei. This would mean that eukaryotic organisms could have appeared much earlier than most scientists would have expected.

MICROFOSSILS FROM 3.5 BILLION YEARS AGO?

The early results from South Africa and Australia set the stage for a major controversy regarding the oldest evidence of life. In the mid-1960s, William (Bill) Schopf was a graduate student at Harvard, working with Elso Barghoorn. After completing his degree, Schopf joined the Geology faculty at the University of California, Los Angeles (UCLA) and began working on the central question of this chapter: What evidence can be discovered that will establish the antiquity of life on Earth? Stanley Awramik, a new faculty member at the University of California, Santa Barbara (UCSB), was also inspired by Barghoorn's approach and began to ask the same question. Awramik worked not only with Barghoorn, but also with Preston Cloud at UCSB, Andrew Knoll at Harvard, and Malcolm Walter in Australia. In 1983, Awramik was first author of a paper in *Precambrian Research* coauthored by Schopf and Walter, which reported possible microfossils in the cherts of Western Australia. Four years later, Bill Schopf and Bonnie Packer coauthored a paper in *Science* that finally caught everyone's attention, in part because *Science* has a much broader readership than *Precambrian Reseach*, but also because the results seemed clear enough to produce a consensus view that bacterial life was present on Earth at least 3.5 billion years ago.

Several images of the apparent microfossils are shown in Figure 4. In a later paper Schopf published in 1993, some fairly bold claims were made, for instance, that 11 different kinds of bacteria could be discerned from the images, and that some of these were oxygen-producing photosynthetic cyanobacteria. Not only was Bill Schopf convinced that the fossils were real, but there was sufficient information to describe them as actual bacterial species of cyanobacteria. Making such a claim about a topic as important as the earliest life attracts immediate critical commentary, and Roger Buick published a cautionary note. It was largely ignored, and for nearly a decade, Schopf's interpretation was accepted, and the images were included in textbooks as representing actual fossils of the earliest known life.

FIGURE 4

Microfossils of ancient bacteria. (A) (left panel) shows a micrograph of fossil microorganisms found in a rock formation 775 million years old from Kazakhstan. The right-hand panel shows a three-dimensional image of the same sample performed by confocal laser scanning microscopy (CLSM).

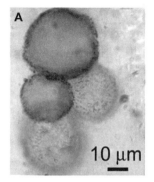
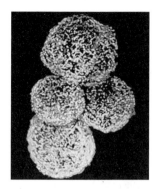

A living cyanobacterial species is shown in the top panel of (B), and the center panel is a light micrograph of an ancient cyanobacterium from the same rock formation as (A). The bottom panel of (B) is a CLSM image of the specimen. The contrast is produced by the organic material remaining in the fossil, which gives off a special kind of light called Raman emission that is characteristic of kerogen, a polymer produced when biological material undergoes aging.

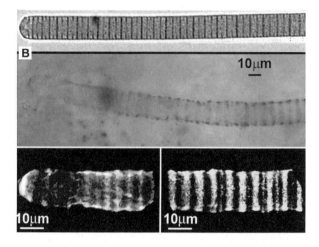

(C) shows a filamentous structure found in 3.5-billion-year-old Australian rocks, four times older than those illustrated in (A) and (B). Is it an actual fossilized bacterium? Probably, because the Raman spectrum is characteristic of kerogen from biological material, and the 2-D Raman image shows interior compartments consistent with cellular life. Images courtesy of William Schopf.

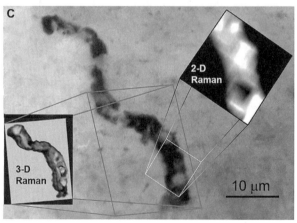

Therefore, it was quite shocking for everyone in the field when a paper from Oxford University appeared in *Nature* in 2002 with Martin Brasier as lead author. In it, the authors challenged Schopf's interpretation of the Australian microfossils. Instead, the authors proposed a less dramatic interpretation—that the apparent fossils were simply artifacts produced when nonbiological carbon compounds form aggregates in hydrothermal vents.

Such a dramatic clash of opinion sometimes arises between personalities that are bold and risk-taking. Bill Schopf is nothing if not bold, and he was quite certain that the evidence was sufficient to make the assertion that these were actual microfossils of the earliest known life. Brasier and his coworkers were bold in their own fashion, and in their judgment, Schopf's evidence was insufficient. Furthermore, they were able to propose and support what is called a null hypothesis. A null hypothesis represents the most likely (and usually least interesting) possible explanation, in this case that the putative microfossils were not produced by life, but by geochemical processing of organic material like that in the Murchison meteorite. Until the null hypothesis is disproven, they argue, it is rash to jump to the conclusion that the dark microstructures are anything more than organic debris.

Nothing is as energizing as having your research results challenged in such a public forum. In order to respond, Bill Schopf has assembled a magnificent microscope facility that can produce three-dimensional images of the microfossils and furthermore provide what are called Raman spectra that tell whether the dark material is simply inorganic graphite or the polymeric kerogen expected to be produced by biological material. During a recent visit to Bill's lab, I sat at the microscope and watched a beautiful image of a microfossil appear on the screen and then rotate to produce a 3-D effect. This was a test run on a fossil from much younger rocks than the Australian Apex chert, but Bill has also applied the technique to the older samples. The resulting Raman results are consistent with the idea that the dark material of the microstructures is of biological origin. The Raman images match the shape of the black microfossil structures, and the spectral pattern is what one would expect of polymerized organic material, not the carbon of graphite (Figure 4C).

The bottom line is that we cannot yet be absolutely certain whether the enigmatic Apex chert structures are artifacts or evidence of the first life. Like jurors, other scientists are waiting for the two sides to present further evidence, which will require new approaches, new samples, and more work. As one of the jurors, what would convince me is more images of microfossils showing a pattern of structures. This is abundantly clear in more recent microfossils from about two billion years ago, but ancient fossils from 3.5 billion years ago are very rare, so this is likely to require an extensive (and very tedious) continuing exploration of ancient rocks. There must also be a convincing geological context. In other words, the fossils should be present in a rock formation that clearly matches the kind of rock that is likely to be a habitat for primitive bacterial colonies. Perhaps most important will be to demonstrate that similar structures cannot be produced by plausible nonbiological starting material.

LIFE 3.8 BILLION YEARS AGO? THE EVIDENCE FROM STABLE ISOTOPES

The case for life existing 3.5 billion years ago would be more convincing if evidence could show that life existed even earlier. Manfred Schidlowski, who worked at the Max Planck Institute in Mainz, Germany, had an idea. He knew that the Isua rocks of Greenland were

3.8 billion years old. Furthermore, even though they had undergone considerable geological processing, the rocks contained detectable traces of organic carbon. Could this carbon be a biomarker for life? The same stable isotope analysis described earlier for analyzing the oxygen content of zircons might provide an answer. The technique takes advantage of the fact that there are two isotopes of carbon, the most common (~99%) being carbon with an atomic weight of 12, mixed with the much less common carbon 13 (~1%). Schidlowski also knew that when living organisms such as photosynthetic bacteria take up carbon dioxide to use as a carbon source for growth, there is a slight tendency to capture the lighter isotope in preference to the heavier isotope. This difference can be compared against a nonbiological carbon compound such as calcium carbonate in limestone. Thousands of such measurements have consistently shown that the organic material produced by living organisms is "lighter" in the sense that it contains more carbon 12 than expected from comparison with the carbon in limestone. For convenience, the difference is expressed in parts per 1,000 (rather than parts per 100, which would be the more familiar percent in common use), and typical results for biomass range from -20 to -50. When Schidlowki measured the same value for Isua carbon it was -27, in the range expected for carbon that had been processed by living organisms. (The minus sign indicates a range lighter than carbonate mineral.)

Could life really be that old? Despite the intrinsic interest of Schidlowski's result, it received little attention because there was still uncertainty about the actual age of the Isua rocks and their geological history. But then in the mid-1990s, Stephen Mojzsis, a graduate student at UCLA, decided to try a new technique on Isua rocks. Mojzsis had access to the kind of ion microprobe instrument that John Valley had used on zircons. Mojzsis reasoned that if the organics really were a product of life, they might be associated with a calcium phosphate mineral called apatite that was scattered throughout the mineral matrix of Isua rocks. Apatite is the mineral of bones and teeth today, and phosphate is used by all living organisms. Perhaps during diagenesis it would be turned into calcium phosphate while the organic carbon content of the microorganism would become graphite particles embedded in the apatite matrix. The apatite mineral would then protect the graphite from diagenesis. (Graphite is a form of carbon produced when organic carbon is subjected to heat and pressure sufficient to drive off all of the other atoms present, particularly hydrogen and oxygen.)

Mojzsis used the microprobe to scan microscopic apatite deposits in rocks from Isua as well as rocks called banded iron formations that were sampled from Akilia, an offshore island in Greenland. By this tim e, the rocks had been dated much more accurately by other workers to be 3.37 billion years for Isua and 3.85 billion years for the Akilia rocks. The results were astonishing: -30 for Isua and -37 for Akilia! This was almost too good to be true because these numbers are well within the range expected if the carbon atoms had been processed by some form of metabolism.

Alas, these results are now also in dispute, not because the measurements are incorrect, but instead due to interpretation of the rocks themselves. Mojzsis assumed, as others had done, that the Isua rocks were primarily sedimentary material produced in an aqueous environment that had later been subjected to high pressure and temperature. In their 2002 paper, Mark van Zuilen and coworkers at the Scripps Institution of Oceanography showed that the rock was not sedimentary, but instead was produced when hot water reacted with older crustal minerals. Carbon in the form of graphite can be produced when carbonate undergoes high temperature processing, rather than from a biological source. Van Zullen's group concluded that the light carbon in the graphite must have been produced by geochemical reactions: "The new observations thus call for a reassessment of previously presented evidence for ancient traces of life in the highly metamorphosed Early Archaean rock record."

WHAT WAS IT LIKE FOUR BILLION YEARS AGO?

We can now put together all the information from Chapters 2 and 3 by describing what Earth would have looked like if we could somehow travel back in time four billion years to the early Archaean Eon, before life began. We might find ourselves standing on the rocky shore of a volcanic island. It is very hot, around 70°C, and the atmosphere is a toxic mixture of carbon dioxide, nitrogen and volcanic gases. For this reason we must wear protective clothing, an air-conditioned space suit with an oxygen supply. In the distance we can see other land masses rising from the sea, some of which are active volcanoes. The rocks beneath our feet are composed of dark lava with volcanic ash filling the crevices. Hot springs boil all around us. The fairly salty sea water has an odd greenish tint from all the dissolved iron that it contains. White deposits of dried salt on the lava rocks show where small tide pools have evaporated. Rain falling in the nearby volcanic peaks causes freshwater ponds to form a few meters above the beach. These are constantly being filled by small rivulets of water cascading down the mountainside, then drying out in the heat. Suddenly the landscape is brilliantly illuminated for several seconds as a blinding white streak silently crosses the sky and descends into the sea just over the horizon, where an even brighter flash of light appears. A minute later we hear a thunderous roar, followed by a pulsing shockwave that nearly knocks us over. A small asteroid, maybe 100 meters in diameter, has penetrated the atmosphere at 20 kilometers per second and has crashed into the ocean several miles away—one of many such impacts that occur every day. A thin dark line on the ocean horizon advances toward us. The tsunami from the meteorite impact is heading our way, so we fast forward four billion years to a more familiar setting.

Back home, we have only one thought from our experience in the prebiotic Archaean Eon: How could life begin in such an unpromising environment? And this brings us to a fundamental question that is still unanswered: Is the origin of life a common event? In that case, there might have been multiple origins of life, but each attempt was snuffed out by the giant impact events associated with the Late Heavy Bombardment. The life we see today are the survivors that managed to live through the last giant impact or that started up as soon as it was safe to do so.

The alternative is that the origin of life is exceedingly rare, requiring half a billion years or more to happen just once on a habitable planet like early Earth that has just the right set of conditions. We just don't know the answer yet, and this book is a progress report on our attempts to find out. If I had to guess, I would be optimistic. I think that we are just at the point of being able to produce a form of synthetic life in the laboratory. When we can do that, we will have a much better understanding of how life can begin, and how likely it is that a similar process could occur on early Earth

CONNECTIONS

The stable isotope composition of oxygen in zircons supports the presence of oceans and land masses at least 4.4 billion years ago. The initial reactions leading to the production of complex organic compounds in bodies of water could have begun at that time. Evidence from stable isotopes of carbon suggests that organic carbon was being processed by living organisms about 3.8 billion years ago, but the evidence for this is under debate and not yet resolved. The first possible microfossils of bacterial life are present in Australian rocks that are 3.5 billion years old, but these are also disputed. The earliest undisputed microfossils are in rocks that are 2.5 billion years old. Bacterial life was abundant more than two billion years ago, and molecular fossils called steranes suggest that eukaryotic life appeared about this same time. Simple forms of multicellular life were present over one billion years ago, and hard-shelled organisms became abundant in the fossil record during the Cambrian Period 540 million years ago.

The next chapter will describe the kinds of organic compounds that were likely to be present in the environment at the time of life's origin, and then we will see how such compounds can interact with each other to produce ever more complex structures on the evolutionary pathway toward the first forms of life. We will see how experiments in the laboratory attempt to simulate conditions on early Earth, but an important point is that none of these simulations comes close to the actual conditions described in Chapters 2 and 3. The reason for this is a good one, having to do with standard scientific practice to keep experimental conditions as simple as possible in order to prevent confusion. Such simplicity, although essential in designing experiments, may also be limiting progress in origins of life research. I will argue that there is another principle I call sufficient complexity that is essential if we are to reproduce the origin of life in the laboratory.

4

CARBON AND THE BUILDING BLOCKS OF LIFE

One approach to understanding life's origin lies in reducing the living cell to its simpler chemical components, the small carbon-based molecules and the structures they form. We can begin by studying relatively simple systems and then work our way up to systems of greater complexity. In such an endeavor, the fascinating new science of emergence points to a promising research strategy.

ROBERT HAZEN, 2005

The first time I ground up a marble-sized sample of the Murchison meteorite, a strange, penetrating odor rose from the mortar, simultaneously smoky, dusty, and sour, reminiscent of a cigar butt or the contents of a vacuum cleaner bag. It was a distinctive smell that I would now recognize anywhere. It was the odor of outer space, nearly five billion years old, the aroma of organic compounds that were delivered directly to Earth during the birth of our solar system. Nothing could be more convincing of the link between cosmic processes and life on Earth.

Carbon compounds present in meteorites and comets also abound in the molecular clouds that give rise to solar systems. Why is it so important for the origin of life that organic compounds are scattered throughout our galaxy? There are only two possible sources for the organic compounds required for the origin of life. If they were present in the molecular cloud from which Earth formed, then they were delivered by a rain of comets and dust particles after the moon-forming event. The alternative source is synthesis by chemical reactions on Earth's surface. We don't yet know which source was primary, so let's examine both possibilities.

There is one thing we know with certainty: All of the carbon atoms present in living organisms were delivered to early Earth during accretion, along with water, nitrogen, sulfur, phosphorus, iron, and silicate minerals that composed the dust particles in the solar nebula. Because accretion was such a violent, high-energy process, at first any organic carbon compounds were broken down into gases, mostly carbon dioxide. But after our planet reached its final mass with oceans and atmosphere present, organic carbon compounds continued to be delivered. We are certain of this. In fact, it is still happening today, although much less frequently than four billion years ago.

ORGANIC CHEMISTRY AND LIFE

First, we need to learn something about the organic compounds that comprise life. The discipline of organic chemistry was founded in the 1800s, when chemists became interested in the chemical compounds associated with life and decided to name their discipline using a word—organic—having to do with life. But as our knowledge of carbon chemistry grew, it became clear that this usage was misleading. There were lots of carbon-containing compounds that had nothing to do with living organisms, so a new discipline called biochemistry was founded to deal specifically with the chemistry of life. Organic chemists could then devote their studies to the carbon compounds used in plastics, oils, dyes, pharmaceuticals, herbicides, pesticides, and thousands of other chemicals.

My own introduction to organic chemistry came during my second year at Duke University, where I was a chemistry major with a slide rule in a leather case dangling from my belt. (I still have the slide rule, now joining the abacus as an antiquated way to multiply and divide. Yet I can astonish my students by doing a multiplication faster on my old slide rule than they can on their programmable calculators.) The introductory organic chemistry course was taught by Lucius Aurelius Bigelow. Professor Bigelow was in his 60s at the time, bald with a fringe of white hair, and needed to wear glasses that vastly magnified his bright blue eyes. He had a passion for his subject, and we watched in amazement three times a week as he covered the blackboard from top to bottom with organic reactions, lecturing entirely from memory. Professor Bigelow's lectures were extremely dry, but the accompanying lab was an adventure. Laboratory safety is paramount these days, but back then it consisted of little more than a warning that ether was flammable. We were allowed to dispose of solvents down the drain, which is unthinkable today. Occasionally a student would dump a mixture of ether and lithium aluminum hydride into the sink. When the lithium compound reacted with water in the drain it ignited the ether, producing an impressive fireball that singed undergraduate eyebrows. We learned laboratory safety the hard way.

In the lab, our first exercise in organic synthesis was to make something called hippuric acid. The result was a white crystalline precipitate with an odor reminiscent of barns and corrals. Hippuric acid is the compound that gives horses their distinctive aroma. (The word "hippopotamus," or river horse, has the same Greek derivation.) The lesson was that virtually all the chemicals of life are organic compounds, and we can often synthesize them using the tools of organic chemistry.

THE GAME OF LIFE

Now we can introduce the main biochemical components of a living cell and ask what chemical and physical properties they have that could tell us something about the origins of life. Although a cell is incredibly complex, a list of its atomic components is fairly brief. Think about the game of checkers. The parts are very simple, just black and red pieces on a checkerboard with 64 squares, and the rules that govern the way the pieces move on the board are also easy to understand. However, the situations that arise during an actual game of checkers are so complex that only in 2007, using the most powerful computer, was the game completely analyzed. (It turns out that if two players make perfect moves, the game always ends in a tie.)

In living organisms, as in checkers, immense complexity arises from the way specific rules govern a few basic pieces. Instead of two colors of checker pieces, life is based on six biogenic elements abbreviated CHONPS. The list is almost but not quite pronounceable, which helps to remember it, and is also approximately in the order of abundance (by weight) of the atoms in the proteins and nucleic acids of life. Carbon (C), phosphorus (P), and sulfur (S) are solids at ordinary temperature, and hydrogen (H), oxygen (O), and nitrogen (N) are gases. These elements comprise more than 99% of the water and organic matter in a living cell. One of the chemical rules related to life processes is that the six biogenic elements combine into four basic kinds of molecules, which, in turn, assemble into the structures that make a cell. This is the reason that CHONPS are called the biogenic elements, because they can be assembled into sets of simple molecules that in turn can be linked to form strands of proteins and nucleic acids. Beginning with H and C, we will now construct those four basic molecules, adding one more biogenic element at each step to show how chemical complexity can emerge from combinations of the six elements.

HYDROGEN AND CARBON

There is a great divide in the physical properties of organic compounds involved in life. Most of the familiar compounds like proteins and nucleic acids tend to be soluble in water, but if we grind up bacteria, or spinach leaves, or an oyster in alcohol we discover a second group of compounds that are also essential for life. The proteins and nucleic acids turn into an insoluble precipitate, but the alcohol dissolves fatty substances called lipids, which are defined by the fact that they are soluble in organic solvents. Examples of lipids include the fat used to store energy, the cholesterol and phospholipids that form cell membranes, and even certain vitamins like A, D and E. Lipids are soluble in organic solvents because they contain long chains composed of hydrogen and carbon, from which we get the word "hydrocarbon." Hydrocarbons are so simple that we can represent them by substituting letters of the alphabet for atoms in the molecule. For instance, the formula for octane, one of the main hydrocarbons of gasoline, is $H_3C-CH_2-CH_2-CH_2-CH_2-CH_2-CH_2-CH_3$, or C_8H_{18} in the usual abbreviation used by chemists. The eight carbons in the chain give octane its name. Other hydrocarbons in this series include pentane, hexane, heptane, nonane, and decane, with 5, 6, 7, 9, and 10 carbons in their chains.

Most of the chemical bonds linking the carbon atoms in hydrocarbons are called single bonds, abbreviated C–C, which are produced by a pair of shared electrons. This is an important point to understand, because it is fundamental to the way that chemistry is related to the origin of life:

The laws of chemistry arise from the fact that chemical structures and chemical reactions involve electrons in the outermost shells of atoms. The processes of metabolism, growth, and reproduction that are characteristic of life are electronic at the most fundamental level.

Keep that in mind as we delve deeper into the way life works, and how life began. We will often refer to electrons, because they are the means by which atoms are linked into molecules. Every chemical reaction involves changes in the electronic structure of the molecules that take part in the reaction. You should also understand that electrons are not permanently bound to a given atom or molecule, but instead jump around during chemical reactions. For instance, electrons are transported between molecules in the reactions by which energy is captured and used by living organisms. The first thing that happens in photosynthesis by green plants is that red light energy is absorbed by a green chlorophyll molecule and causes its electronic structure to jump from what is called a ground state to an excited state with extra energy content. As a result, an electron can pop off the chlorophyll molecule. By a complicated series of reactions, that electron ends up on carbon dioxide, which ultimately is incorporated into carbohydrates and fats that are used for energy by the plant itself and the animals that eat plants, including us. As you read this, in every cell in your body, hundreds of tiny organelles called mitochondria are grabbing electrons from metabolic products of fats and carbohydrates and delivering them to oxygen. The result is a voltage of about 0.2 volts across the mitochondrial membrane, and this is the energy used to synthesize ATP, the ultimate energy source of all life. This process will be discussed in more detail in Chapter 6, when we consider how the first forms of life used energy to drive synthetic reactions required for growth.

Hydrocarbons are a good way to introduce the fundamental laws of chemical bonds that are common to life's molecules. As noted earlier, a chemical bond is produced when pairs of electrons are shared between two atoms. The simplest of all chemical reactions occurs when a hydrogen atom, composed of one proton and one electron, meets another hydrogen atom. The two atoms instantly form a chemical bond by pairing up their electrons to become hydrogen gas. The resulting product can be abbreviated in a number of ways, such as H:H to indicate the pair of electrons or H–H where it is understood that the line represents the electron pair or simply H_2. Other familiar gases do the same pairing: Oxygen and nitrogen in our atmosphere are not present as individual atoms, but instead as O_2 and N_2. The electronic structure of the bonds in these gases is more complicated than a simple pair of electrons, but the details are not needed at this point. Just remember that atoms form molecules by sharing pairs of electrons.

Hydrocarbons containing only single bonds are referred to as saturated hydrocarbons. Other hydrocarbons use two pairs of electrons to form double bonds, abbreviated C=C, and are referred to as unsaturated. Sometimes hydrocarbons form rings, and then they are called cyclic hydrocarbons. An example is cyclohexane (C_6H_{12}). But a cyclic compound with six carbons containing three double bonds in the ring is called benzene, an aromatic hydrocarbon. Cyclohexane doesn't have a strong aroma, but benzene certainly does, and so do other aromatic hydrocarbons like naphthalene.

Cyclic hydrocarbons are not limited to just one ring, but can have two or more rings linked together. These compounds have an obvious descriptive term: polycyclic. If the compounds also have double bonds as in benzene, they are polycyclic aromatic hydrocarbons (PAHs), which turn out to be the most abundant organic compounds in the universe. PAH compounds and their derivatives contribute to the aroma of smoke, hot tar, diesel fumes from truck engines, and the Murchison meteorite.

Despite being common in interstellar space, few PAH derivatives are involved in life processes. One exception is naphthalene, formerly used to repel the moths that eat holes in wool clothing stored in closets. Naphthalene is simply two benzene rings joined together along one edge, and its derivative naphthoquinone is part of the electron transport chain in mitochondria. Other PAH are, in fact, dangerous, such as the carcinogenic benzpyrene of cigarette smoke.

CARBON, HYDROGEN, AND OXYGEN

Now that we have described the primary chemical bonds, the rest of the major molecules of life can be introduced. We can begin by adding oxygen to hydrocarbon chains. Life does not use pure hydrocarbons like octane, but instead incorporates longer chains of hydrocarbons that have a chemical group resembling carbon dioxide at one end. This oxygen-containing chemical group is –COOH, dubbed carboxyl because it contains both carbon and oxygen atoms. Hydrocarbon chains with a –COOH group at one end have two special properties. Most important is that the carboxyl group provides a kind of chemical handle that can be used to link a hydrocarbon chain to other molecules. For instance, in the triglycerides we call fat or vegetable oil, three of the chains are linked to glycerol through their carboxyl groups. The second property of the carboxyl group gives a name to these special molecules. When exposed to water, the –COOH can lose the H in the following reaction:

$$-COOH \ \{\Rightarrow\} \ -COO^- + H^+$$

Whenever a chemical can release a hydrogen ion (also called a proton) that chemical is referred to as an acid. The –COOH is therefore called carboxylic acid, and hydrocarbon derivatives with a carboxylic acid group are called fatty acids, because they are primary components of fat. At alkaline pH ranges, fatty acids lose the hydrogen ion and become negatively charged, and then they are called soaps, the familiar compounds of everyday life.

Molecules resembling fatty acids likely played a crucial role in the origin of life, because the functional molecules of living cells must be maintained in a compartment of some sort. The physical properties of fatty acids and their derivatives allow such molecules to assemble into the membranous container surrounding all cells. Without such a container, cellular life could not exist.

Carbon, hydrogen, and oxygen are also the elemental components of carbohydrates. The word comes from the fact that hydrogen and oxygen are present in a typical carbohydrate in a ratio of two to one, approximating the ratio of water, so that a carbohydrate is literally "watery carbon." An example is glucose, or "blood sugar" ($C_6H_{12}O_6$), which can polymerize to form starch and cellulose. (The word "sugar" commonly refers to sucrose, the sugar we buy in stores, but in biochemistry it also applies to simple carbohydrates like glucose and fructose.) Other important examples are the sugars ribose and deoxyribose, which give ribonucleic acid (RNA) and deoxyribonucleic acid (DNA) their names. When linked together with phosphate into chains, these sugars form the backbone of nucleic acids, the central information-processing molecule used by all life today.

Carbohydrates are a ubiquitous energy source for life. For instance, the metabolic process called glycolysis releases chemical energy and is present in the cells of every living organism from bacteria to the human body. Glycolysis consists of a series of enzyme-catalyzed reactions in which glucose, with six carbons, is broken down into smaller molecules with three carbons. ("Glycolysis" comes from words meaning "sugar breaking.") The ubiquity of glycolysis in life processes suggests that carbohydrate metabolism has ancient roots, perhaps evolving in the earliest forms of life. Carbohydrates, however, have only half the energy content of hydrocarbon chains, which is the reason that most organisms store energy in the hydrocarbon chains of fat rather than as sugars.

HYDROGEN, CARBON, OXYGEN, AND NITROGEN (AND A LITTLE SULFUR)

CHON(S) are the primary elements of amino acids. Examples include glycine ($C_2H_6O_2N$) and alanine ($C_3H_9O_2N$), two of the 20 biological amino acids. Amino acids take their name from the fact that they have both an amine ($-NH_2$) and an acid group ($-COOH$). These two chemical groups are fundamental to all life because they permit amino acids to polymerize into proteins, the primary polymers of living systems. Two amino acids— methionine and cysteine—contain sulfur (S), and this is why sulfur is included among the biogenic elements.

The most important thing to understand about amino acids is that most cell structures are composed of proteins. The origin of proteins and their relation to nucleic acids represents a huge gap in our understanding of how life began. In other words, an important clue to the origins of life is the process by which amino acids can combine into primitive protein-like polymers in the absence of pre-existing life. In later chapters, we will discuss several ways that this could occur.

HYDROGEN, CARBON, OXYGEN, NITROGEN, AND PHOSPHOROUS

CHONP are the elements of nucleotides, such as adenosine monophosphate and adenosine triphosphate. Five different nucleotides polymerize to form nucleic acids (DNA and RNA), the information-processing molecules of the cell, and adenosine triphosphate

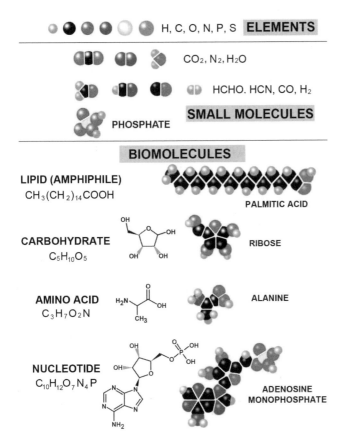

FIGURE 5

Six biogenic elements make up the four major molecular species used by all forms of life. In this figure, the elements are coded in shades of gray according to the scheme in the top row of elements. It is certain that the small molecules shown in the center group were available in the early Earth environment. Some, such as formaldehyde (HCHO) and cyanide (HCN), readily react to form larger molecules such as amino acids and nucleobases. The molecules in the lower group are representative of the four primary classes of biomolecules involved in life processes.

(ATP) serves as the primary energy currency of life by transporting chemical energy throughout cells to energy-dependent processes such as muscle contraction, metabolism, and nervous activity.

The molecular structures of life have their own kind of beauty, and examples of the four main molecular species are summarized in Figure 5. The CHONPS atoms are shown in different shades of gray to indicate how the biogenic elements are combined to produce the pieces of the game of life.

INORGANIC IONS ARE ALSO ESSENTIAL FOR LIFE

This chapter is about the chemistry of life, mostly organic chemistry, but it is a good place to introduce the inorganic aspect of living systems. At the cellular level, this is mostly related to ions in solution, but it also includes metal atoms like iron, copper, magnesium, and zinc that are incorporated into the chemical structure of certain proteins. The principal ions in solution that are important for life include sodium, potassium, magnesium, calcium, and hydrogen ions, which are all positively charged and are referred to as cations. Negatively charged ions are called anions, and the most important ones are chloride, carbonate, phosphate, and sulfate.

The electrical charges on ions play crucial roles in life processes, so we need to understand how something in solution becomes electrically charged. Except for hydrogen ions, the cations listed above are all metals. I have sodium metal in my lab, and when I give undergraduate lectures in chemistry, I bring some to class. Metallic sodium is very soft, so it is possible to cut it with a knife. I carefully cut off a small piece about the size of a pea and then drop it into a fish bowl with water in the bottom. The most amazing thing happens. The sodium metal immediately melts and begins to whiz around on the surface, giving off sparks. A few seconds later it explodes with a pop and a puff of white smoke.

What happens is that the sodium metal reacts with the water and turns into sodium ions:

$$Na \text{ (metal)} + H_2O \{\Rightarrow\} Na^+ \text{ (sodium ion)} + OH^- \text{ (hydroxide anion)} + H \text{ (hydrogen)}$$

So much energy is released by the reaction that it melts the sodium metal. The H that is produced immediately combines to produce H_2, or hydrogen gas, and that is what produces sparks and the explosion when it ignites in the air.

Metals like sodium have electrons that they can give away to water in a reaction, and when they lose negative electrons the metal atoms become positively charged cations. Sodium and potassium ions carry one positive charge, while magnesium and calcium ions have two charges. All of these metals are present as ions in sea water because they were delivered during the primary accretion phase of Earth's formation.

The anions important for life are a little more complicated because they are produced when a central nonmetallic atom reacts with oxygen. The resulting compound has one or more extra electrons and is therefore negatively charged. Oxidized carbon produces bicarbonate $(HCO_3)^-$ and carbonate $(CO_3)^{2-}$ with one and two negative charges, sulfur becomes sulfate $(SO_4)^{2-}$ with two negative charges, and phosphorus becomes phosphate $(PO_4)^{3-}$ with three negative charges. Chlorine can also react with oxygen, but it is mostly present in oceans simply as chloride anion $(Cl)^-$ with one negative charge.

One last pair of ions is essential to understand: cationic hydrogen ions and anionic hydroxide ions. Hydrogen is the smallest element, consisting of a single proton with a single electron. Hydrogen ions are produced whenever hydrogen loses the electron, leaving behind the nucleus, so we often refer to them as protons (H^+). A hydroxide anion is simply OH^-.

Water is continuously dissociating into hydrogen and hydroxide ions in the following reaction:

$$HOH \{\Leftarrow\}\{\Rightarrow\} H^+ + OH^-$$

At neutral pH 7, only one water molecule in 550 million is dissociated at any given time, and the concentrations of H^+ and OH^- are equal: 10^{-7} M. But at acidic pH ranges, protons cause the effect called acidity; and at higher pH ranges, hydroxide ions produce alkalinity. The reason this is important is that protons, hydroxide ions, and other common inorganic ions are involved in virtually every aspect of life processes. Furthermore, the most common cations and anions react to produce minerals such as limestone (calcium carbonate, or $CaCO_3$), gypsum (calcium sulfate, or $CaSO_4$), and the calcium phosphate mineral called apatite, all of which played roles in the origin of life. Silicon never got caught up in living systems, but its oxidized anionic compound called silicate (SiO_2) is the primary anionic component of all the silicate minerals such as the basaltic lava that formed the most common land masses on early Earth.

WHERE DID ORGANIC CARBON COME FROM?

Two of my scientific colleagues—Bill Irvine and Lou Allamandola—introduced me to a fundamental yet little known fact of life: *We live in an organic universe.* Most of the carbon drifting as immense dust clouds in our galaxy exists as organic compounds. In Chapter 1, we learned that the elements of life—carbon, oxygen, nitrogen, phosphorus, and sulfur—are produced by stars as they approach ultra-hot temperatures toward the end of their lifetime and then are dispersed into the interstellar medium when the star explodes. But elements by themselves have no role in the structure of living organisms. Even though we give great emphasis to carbon, the fact is that carbon by itself is just the black soot in smoke, the graphite in pencil lead, or the diamond in a wedding ring. Nitrogen, oxygen, and hydrogen are just colorless gases. Only when carbon combines with hydrogen, oxygen, and nitrogen do we have anything of biological relevance.

We know from infrared and radio astronomy that much of the carbon drifting in interstellar space is organic. Really cold matter in molecular clouds emits radiation in the microwave region, and each chemical bond in a compound produces a specific wavelength. If the radiation is detected with a radio telescope, it is possible to decipher the kinds of bonds present and determine the nature of the compound. Thirty years ago, when I first heard Bill Irvine speak at a Gordon Conference on the Origin of Life, I naively wondered why a radio astronomer would be invited. But then, as he began to show his slides, the mystery was solved. Bill presented clear evidence that dense molecular clouds, the nurseries of stars and solar systems, contain nearly 100 species of organic compounds. (See Table 1 for a list of the more prominent compounds.) Some of these are familiar, such as cyanide, formaldehyde, methanol, ethanol, formic acid (named after *formica*, Latin for "ants," which release formic acid as a caustic spray when disturbed), and acetic acid (the sour component of vinegar). Other compounds were exotic, including a chain of nine carbons with a nitrogen atom at one end. Such a weird compound could not exist on Earth, but in the cold of outer space, it is stable. The connection between radio astronomy and the origins of life became obvious as I listened to Bill. If molecular clouds gave rise to stars, planets, and solar systems, maybe some of the organic matter in the clouds was delivered to early Earth four billion years ago to help life get started.

TABLE I   Examples of Molecules Detected in Interstellar
Molecular Clouds

Number of Atoms	Molecules
2 atoms	carbon monoxide (CO)
3 atoms	water (H_2O), carbon dioxide (CO_2), hydrogen sulfide (H_2S), hydrogen cyanide (HCN)
4 atoms	acetylene (C_2H_2), ammonia (NH_3)
5 atoms	methane (CH_4), formic acid (HCOOH)
6 atoms	methylamine (CH_3NH_2), methanol (CH_3OH)
8 atoms	acetic acid (CH_3COOH)
9 atoms	dimethyl ether [$(CH_3)_2O$], ethanol (CH_3CH_2OH)
10 atoms	glycine (NH_2CH_2COOH)

We also use infrared astronomy to learn about the chemistry of interstellar space, and this is Lou Allamandola's research specialty. Organic compounds, diffusing as gases in the vacuum of space, absorb light in the infrared region. We can't see infrared energy, but sensitive instruments can measure the spectrum, which gives us information about the source. For instance, if an organic substance absorbs infrared light at a frequency of 2,900 waves per centimeter, we can be pretty sure that the infrared light is absorbed by C–H bonds in a hydrocarbon. Allamandola and his colleagues at NASA Ames observed the spectrum of light from stars that passed through the edges of molecular clouds, and they discovered a strange infrared absorbance peak with a wavelength of 3.4 micrometers. In the laboratory, this peak could be produced by light passing through a gaseous mixture of polycyclic aromatic hydrocarbons (PAHs) described earlier in this chapter. When the amount of PAH required to produce the 3.4-micrometer peak is calculated, the astonishing result is that PAHs represent most of the carbon present in interstellar space.

PREBIOTIC MIXTURES OF ORGANIC COMPOUNDS

Now we need to learn more about meteorites in order to understand the significance of their organic content. Most meteoritic material comes to Earth as dust particles. Something like 30,000 tons of dust particles fall into Earth's atmosphere every year, and larger particles the size of sand grains can be seen in the form of meteor showers such as the Perseids, Orionids, Leonids, and Geminids. These showers are named after the constellation from which they appear to emanate in a radial pattern, but if you could watch them from a high-flying aircraft, you would see that this is an illusion of perspective because they actually fall straight down through the atmosphere. Most of these "shooting stars" are particles shed by comets hundreds to thousands of years ago, and we see them when the orbit of Earth intersects the orbit of the comet and passes through the remains of its trail. More rarely, larger chunks of stone and metal enter the atmosphere and produce bolides or "fireballs." Every few years, one of these boulder-sized objects survives its fiery journey through the atmosphere, and fragments fall to Earth as meteorites. The Murchison meteorite described in Chapter 1 is an example of such a fall, and we now know that such objects are produced by collisions in the asteroid belt between Mars and Jupiter.

We get our impressions of meteorites by visiting museums and seeing lumps of metal that look as though their surfaces have been melted by a blow torch. These are iron-nickel meteorites, and they get into museums because a heavy lump of iron is so different from other sorts of rocks we might run across while hiking. But metallic meteorites are relatively rare. Most meteorites are composed of silicate minerals and resemble any other rock you might find at the beach. There is one place on Earth where we can pick up virtually all kinds of meteorites that have fallen and compare them. When meteorites fall in Antarctica, they tend to get concentrated in certain ice fields and can be easily recognized as dark objects against the white ice. Since 1969, more than 10,000 specimens have been collected by American and Japanese scientists. Of these meteorites, it can be calculated that approximately 4% are nickel-iron, and 85% are stony meteorites called chondrites. About 5% of the chondrites are carbonaceous meteorites. Figure 6 shows the major organic carbon compounds that have been detected in the meteorite that fell near the town of Murchison in southeastern Australia. The compounds in this mixture represent a plausible guide to the kinds of organics available on prebiotic Earth because they would either have been delivered during Late Heavy Bombardment described in Chapter 3, or they would have been synthesized by the same kinds of reactions that went on in the Murchison parent body, a kilometer-sized asteroid that was hot enough for a while to melt its ice content and make liquid water. As you can see in Figure 6, more than 90% of the organic material is a kerogen-like polymer composed of polycyclic aromatic compounds linked together by a variety of chemical bonds. The word "kerogen" comes from the Greek word for "wax," and refers to organic material in sedimentary rocks that can be distilled to produce a light oil, sometimes called kerosene. The kerogen-like polymer of meteorites would not be very useful for life processes unless it was somehow broken down into simpler chemical compounds.

The last 0.5% of the Murchison meteorite contains all the biologically relevant compounds—those that are soluble in water and organic solvents. These include more than 70 amino acids, six of which are components of proteins: glycine, alanine, valine, aspartic acid, glutamic acid, and proline. There is also a suite of acidic compounds ranging from

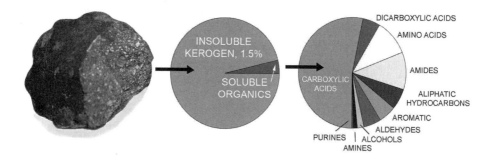

FIGURE 6

Organic compounds in the Murchison meteorite. The organic composition of the meteorite is about 1.5% percent of the total mass, and is mostly present as an insoluble kerogen-like material. A smaller fraction contains thousands of compounds that are soluble either in water or organic solvents, including more than 70 species of amino acids and small amounts of amphiphilic material in the form of mixed carboxylic acids. Even smaller amounts of purines, pyrimidines, and simple carbohydrates have also been detected.

acetic acid with two carbons, all the way up to soaplike acids with a dozen carbon atoms in the chain. Some carbohydrates are present, along with a small amount of adenine, one of the bases of nucleic acids. The most important take-home message is that if these compounds were present in the original asteroids from which meteorites are derived, it makes it plausible that a similar set of compounds would have been available on early Earth.

COMETS ALSO CONTAIN ORGANIC COMPOUNDS

Comets have been called "dirty snowballs" because they are composed of 40% ice, 20% silicate mineral dust, 5% CO_2 and CO, and an estimated 20% organic compounds. Much of the organic material seems to be a black polymer similar to the kerogen of carbonaceous meteorites, but there are also reactive molecules like methanol, formaldehyde, methane, and even a small amount of hydrogen cyanide. The evidence for this composition is based in part on data from NASA's Deep Impact experiment, in which an 800-pound metal probe collided with the comet Tempel 1 in 2005. The probe was traveling at 30,000 miles per hour when it hit, and the energy of the impact sent many tons of comet substance into the surrounding space. The vapor cloud was then analyzed by a technique called infrared spectroscopy. Every organic compound absorbs specific wavelengths of infrared light, and the patterns of absorbed light indicate what kinds of molecules are present.

Where did all this stuff come from? The best guess is that it was synthesized in the molecular cloud that produced our solar system, in which dust particles aggregate into asteroids and comets in the early stages of planet formation. What is certain is that comets delivered a substantial fraction of the water in Earth's oceans.

Even though most of the organic content would be destroyed by the energy of the impact, simulation experiments have demonstrated that a small fraction was likely to survive.

Now I need to give a perspective on the actual amounts of organic carbon that could have been made available to Earth's surface during late accretion. The very large numbers won't mean much by themselves, but an easy way to turn the numbers into something understandable is to calculate the thickness of the layer of organic material that would have been delivered if that many kilograms were spread out over the entire surface of Earth. First, for comparison, let's calculate the amount of organic compounds now present in all life today. This is mostly in the form of cellulose of plant life, and it is not hard to estimate the total cellulose in forests and prairies. The number comes to 1.8×10^{15} kilograms. Earth has a surface area of about 510 million square kilometers. Imagine that we could somehow spread all of the organic material of life over the surface, including the oceans. It would only be a few millimeters thick, about the height of this square: □.

The next step is to calculate the amount of organic material that was produced by plant life in the Carboniferous Period 300 million years ago and now present as coal, oil, and hydrocarbon gas. This number is also pretty well known, because we keep careful track of the organic carbon compounds that we call fossil fuels. Coal and oil add up to about 10^{15} kilograms, and when oil sands and shales are added, the total buried organic material left behind by preexisting life is approximately 10^{16} kilograms. If this were spread out over Earth's surface it would make a layer 2 centimeters deep (about 1 inch).

Let's compare this with the estimated amount of organic compounds that would arrive with comets during the first billion years of Earth's history. Using the craters on the moon to estimate the rate at which comets impacted early Earth, and knowing their organic content, astronomers Chris Chyba and Carl Sagan estimated that up to 10^{18} kilograms could have been delivered over a period of 100 million years, equivalent to a layer of organic material more than 200 centimeters thick! This would seem to be more than sufficient to supply the organic compounds required for life to begin, but keep in mind that it did not appear all at once. Instead, it was more like the thickness of a soap bubble film arriving every year. Some of it would settle onto early volcanic land masses, some would simply disperse into the ocean to produce a very dilute organic solution of the water soluble components, but most of it would be turned into carbon dioxide and disappear into sedimentary calcium carbonate, otherwise known as limestone. In fact, if we add up all the known limestone mineral and could spread it over Earth's surface, it would make a layer 200 meters thick. This represents the amount of primordial carbon dioxide that was originally present in the atmosphere, but because Earth has oceans, it could turn into sedimentary calcium carbonate minerals and give life a chance to evolve beyond simple bacteria. The bottom line is that a certain amount of organic material would have been delivered during accretion, but we don't yet know how much of it would have survived to participate in the origin of life.

PREBIOTIC SIMULATION EXPERIMENTS SHOW THAT ORGANIC COMPOUNDS CAN BE SYNTHESIZED ON EARLY EARTH

Not all the organic compounds required for the origin of life were delivered to Earth on comets and dust particles. We can be reasonably confident that certain compounds were also synthesized in the atmosphere, hydrosphere, and in volcanic conditions. This brings us to the important topic of prebiotic simulations, which represent the only evidence we have for that claim. In a simulation, we make a set of assumptions about local conditions on early Earth, and then we reproduce those conditions in the laboratory and run experiments to see what happens. The results of the experiment then suggest plausible explanations related to the origins of life, or they rule out implausible ideas. Plausibility, of course, is a judgment call on the part of the experimenter, so the interpretation of simulations can lead to controversy. The fact is that the approach to the origins of life that I am describing in this book will be controversial. Some of my colleagues will find it to be reasonable, while others will reject it because it does not fit their own set of assumptions. I think the best approach is to be open-minded about all possible scenarios and to use our own best judgment in choosing what we believe is plausible. We are exploring a wilderness, but no one yet knows which trail will lead to understanding life's beginning. Each experiment, whether a failure or a success, brings us increased knowledge of the terrain, and at some point, one of the intrepid investigators will happen to find a major clue. Several important examples of prebiotic simulations have already taught us something about the possible factors leading to the origin of life.

ATMOSPHERIC SIMULATIONS

One of the first atmospheric simulation experiments transformed origins of life research from speculation to solid experimental science. In the early 1950s, a young graduate student named Stanley Miller began his PhD research at the University of Chicago, guided by Harold Urey, his advisor. Urey had already won a Nobel Prize for his discovery of the hydrogen isotope called deuterium. He was also first to isolate deuterium oxide, or "heavy water" (D_2O). I have a bottle of this in the lab, and occasionally need to weigh some of it for an experiment. One cubic centimeter of ordinary H_2O weighs exactly 1.0 gram, but the same volume of D_2O weighs 1.1 grams. It really is heavy water due to the extra neutron that is present in the deuterium nucleus along with a proton.

Miller and Urey decided to make a chemical model of the primitive atmosphere of Earth. Urey knew that the outer planets were very high in hydrogen content, along with water, methane, and ammonia, and he reasoned that Earth would have had a similar atmosphere just after it completed the process of planet formation. Taking this to heart, Miller decided to simulate these conditions in the laboratory by enclosing a mixture of gases in a large round flask. As a chemist, he knew that nothing would happen unless

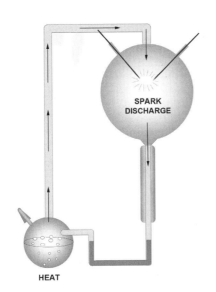
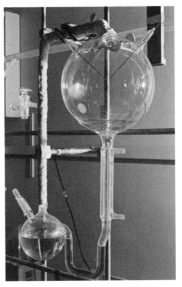

The Miller-Urey experiment was designed to determine whether complex organic compounds could have been synthesized on prebiotic Earth. A mixture of simple gases was subjected to an electrical discharge, and any products of the reaction accumulated in the boiling water in the lower flask. After a few days, a brown material accumulated on the walls of the flask and in the water. When this was analyzed, it was found to contain a mixture of organic compounds, including several amino acids. Photo courtesy of Scripps Institution of Oceanography Archives, UC San Diego.

some form of energy was driving the reactions, so he chose to use an electrical spark to simulate lightning strikes (Figure 7).

The results were spectacular, and even today remain a touchstone for research on the origins of life. Urey himself was skeptical about the experiment and expected the products to be a tarry mess composed of thousands of compounds. He was partly right about the thousands, but the unexpected result was that several of these compounds appeared as distinct spots when separated on paper by chromatography. Several purple and blue spots appeared on the paper when it was sprayed with ninhydrin dye and heated. The spots were amino acids, the fundamental monomers that compose proteins.

Miller went on to propose that the amino acids were synthesized by a previously known reaction called the Strecker synthesis. This reaction was published in 1850 by Albert Strecker, who discovered that cyanide (HCN) and formaldehyde (HCHO) mixtures reacted to produce the amino acid glycine as a major product. In the Miller experiment, the electrical spark caused the methane and ammonia in the gas mixture to form cyanide and a variety of aldehydes, including formaldehyde. These, in turn, reacted to produce amino acids such as glycine and alanine, and a mixture of other carbon

Amino Acids	Percent Yield	Carboxylic Acids	Percent Yield
Glycine	2.1	Formic acid	3.9
Alanine	1.7	Acetic acid	0.5
beta-alanine	0.8	Propionic acid	0.6
Sarcosine	0.3	Glycolic acid	1.9
Aminobutyric acid	0.3	Lactic acid	1.8
Glutamic acid	0.03		
Aspartic acid	0.03		

compounds shown in Table 2. The percent yield is based on the amount of methane originally present in the mixture.

Miller's paper caused a sensation when it was published in *Science* in 1953. It showed that the laws of chemistry allow amino acids to be synthesized under prebiotic conditions, and it seemed that we might be very close to understanding how life began. More than 50 years later, for a variety of reasons that will become apparent, we still don't know how life began, but we do know that most of the simpler organic compounds used by living systems can be synthesized under conditions that could conceivably be present on early Earth.

COMPONENTS OF NUCLEIC ACIDS

After the Miller paper broke the dam, a great deal of effort went into determining how other monomers of life might be synthesized under prebiotic conditions. Along with protein, the other major polymer of life is nucleic acid, composed of chains of a sugar (ribose for RNA, deoxyribose for DNA) and phosphate, with purine bases (adenine, guanine) or pyrimidine bases (cytosine and uracil in RNA, cytosine and thymine in DNA) attached to the sugar groups. The purines and pyrimidines are called heterocyclic, because they contain both carbon and nitrogen atoms linked together in simple ring structures, two rings for purines, one ring for pyrimidines. Phosphate is a component of certain minerals, so it is a reasonable assumption that small amounts of phosphate were available in the prebiotic environment, but what about sugars and bases? As far back as 1861, Alexander Butlerow showed that a mixture of formaldehyde and calcium hydroxide reacted to produce hundreds of simple carbohydrates ranging from three to six or more carbons. Butlerow's observation, now referred to as the formose reaction, was adopted by researchers in the 1960s to explain how carbohydrates might have become components of the prebiotic soup. The fact that hundreds of products are produced is also a problem, because ribose, the sugar required to produce RNA, is only a minor component. However, Steve Benner, at the Foundation for Applied Molecular Evolution, recently discovered that the inclusion of borate in the mix dramatically enhanced the synthesis of ribose and that borate minerals (the source of the borax powder used in laundry detergents) are fairly common in the natural environment. This kind of pleasant surprise is constantly popping up in origins of life research, like the tiny flakes of gold in a prospector's pan that hint at a mother lode somewhere upstream.

In 1961, another chemical simulation produced a major surprise. John Oro, a talented organic chemist who had recently moved to the University of Houston, was interested in cyanide chemistry. Why cyanide? This brings us to the last kind of chemical bond, called triple bonds, which are typically very reactive. For instance, acetylene gas is simply two hydrogen and two carbon atoms held together by a triple bond, and the gas releases large amounts of heat energy when it combines with oxygen, as in acetylene torches used for welding. The point is that hydrogen cyanide is also highly reactive because it has a triple bond between the carbon and nitrogen. Oro found that under alkaline conditions in the presence of ammonia (NH_3), cyanide polymerized according to the following reaction:

$$5 \; HCN \; \{\Rightarrow\} \; H_5C_5N_5$$

The surprise was that $H_5C_5N_5$ is in fact adenine, one of the four bases of DNA and RNA, and present in all life as the AMP (adenosine monophosphate) of nucleic acids and as ATP, or adenosine triphosphate, which supplies energy for most cellular functions. Once again, the rules of chemistry led to the synthesis of a primary component of life. Such discoveries gave increasing confidence to the expectation that the origin of life could be understood. Because the laws of chemistry and physics are universal, the discovery also made it increasingly plausible that life could arise on any planet that has liquid water, a source of energy, and organic carbon compounds. This is why it is so exciting that we now have clear evidence that our sister planet Mars once had shallow seas. It may be possible to discover evidence for past life there, or even existing life deep beneath the surface, where there may still be a source of heat energy and liquid water.

I do want to insert a measured dose of skepticism at this point. We should not necessarily conclude that sugars and the bases of nucleic acids were in fact synthesized this way, or that there were great heaps piled everywhere on early Earth ready to react. Both of the above reactions require concentrated reactants and alkaline pH ranges to produce high yields, but the pH of water associated with volcanic environments on early Earth was probably on the acidic side due to the presence of atmospheric carbon dioxide, a weak acid, as well as sulfur compounds that are also acidic. Yet it is not too much of a stretch to think that small amounts of sugars and nucleobases were present. For instance, traces of adenine have been found in the Murchison meteorite ($\sim$1 ppm, or parts per million, equivalent to 1 microgram per gram of meteorite), and small amounts of simple sugars have also been detected. The fact that such compounds were synthesized in the presolar nebula and asteroids from which carbonaceous meteorites are derived adds weight to the argument that similar reactions could have occurred on young Earth.

GEOTHERMAL SIMULATIONS

John Oro had an appetite for a broader perspective on the chemistry of life's origin. In a pioneering yet often overlooked experiment, Oro and his students passed hydrogen and carbon monoxide gas over hot powdered metal made from a fragment of the Canyon Diablo meteorite. His aim was to demonstrate that another well-known reaction called Fischer-Tropsch type (FTT) synthesis could be catalyzed by the iron-nickel content of an actual meteorite, thus making it more plausible that such reactions could take place on early Earth. In the FTT synthesis, the CO gas molecules transiently adhere to the hot metal surface, which keeps them close enough together so that they can form linear hydrocarbon chains when they react with the hydrogen gas. When Oro analyzed the resulting vapor, it contained a mixture of alkanes, alcohols, and acids that varied with temperature and pressure. A few years later, Ed Anders, working at the University of Chicago, analyzed extracts of the Murchison meteorite and found a variety of hydrocarbons. In a controversial paper published in 1983, Anders and his coworkers suggested that these were products of FTT reactions that occurred on the asteroid parent body of the Murchison.

Well, if hydrocarbons could be synthesized so easily, maybe they were also being synthesized in the volcanic conditions of early Earth! What are the simplest conditions under which hydrocarbons can be synthesized? This question was asked by Bernd Simoneit and his students at Oregon State University, but instead of passing carbon monoxide and hydrogen gas over hot iron, they sealed formic acid or oxalic acid in a small stainless steel pressure chamber and heated it up. Formic and oxalic acid are just one step up the chemical scale from carbon dioxide. If we type carbon dioxide (CO_2) as O=C=O, then formic acid is H–COOH, where –COOH is the acidic carboxyl group. Oxalic acid is two carboxyl groups linked together (HOOC–COOH). As mentioned earlier, formic acid gets its name from the Latin word for "ants" (*formica*), which use it as a defensive agent. And if I chew on the stem of the sorrel (Latin name *Oxalis*) that grows in our weedy back yard, I can taste the pleasantly sour oxalic acid that gives its name to the plant. The word "oxygen," by the way, also has a similar derivation. "Oxy-gen" means something that generates acid. If oxygen reacts with elemental carbon, sulfur, and phosphorus, the result is CO_2, which produces H_2CO_3 (carbonic acid), H_2SO_4 (sulfuric acid) and H_3PO_4 (phosphoric acid). When these compounds dissolve in water, they produce acidic solutions.

The trick used by Simoneit is that oxalic acid heated to oven temperatures breaks apart into hydrogen gas (H_2), carbon dioxide (CO_2), and some carbon monoxide (CO), which are the starting materials for the Fischer-Tropsch reaction. The idea is to simulate the temperature and pressure in a volcano, where the gases can undergo synthetic reactions. Under these conditions, Simoneit and his students have shown that an amazing mixture of fatty acids and alcohols is produced. From such experiments, and from analysis of the organic compounds in meteorites, it seems very plausible that hydrocarbons were relatively abundant in the prebiotic environment, perhaps even forming oil slicks on the ocean surface and washing up on beaches. Hydrocarbons are, by far, the most stable of the organic carbon compounds required by life. For instance, the oil we call fossil fuel is several hundred million years old, and persists long after the proteins, carbohydrates, and nucleic acids of the original organisms are degraded by heat and pressure. Later I will describe how mixtures of fatty acids and alcohols readily assemble into cell-sized membranous vesicles that would have been essential compartments for the origin of life.

MINERAL INTERFACE SIMULATIONS

In 1988, Gunther Wächtershäuser published a remarkable idea that excited tremendous interest, even being featured in *Scientific American*. His idea ran counter to prevailing ones about the origins of life and suggested new experimental approaches involving mineral interfaces. Wächtershäuser is a patent lawyer in Munich, Germany, who enjoys fabricating intricate and novel approaches to the origins of life and then challenging others to test them. He is greatly influenced by the philosopher Karl Popper, who made the point that explanations are useless unless they are falsifiable. In other words, don't publish an idea (or hypothesis) unless you can think of a critical experiment that might prove it to be wrong. This is just the opposite mind-set of an experimental approach that tries to prove an idea is correct.

Wächtershäuser basically turned the entire concept of life's beginning upside down by proposing that life did not arise by assembling pre-existing organic compounds, no matter what the source. Instead, he argued, life began as two-dimensional synthetic chemistry on a mineral surface called pyrite, or sometimes fool's gold, a crystalline mineral composed of iron sulfide. According to Wächtershäuser's idea, pyrite has two special properties. The first is that it has to have a positive surface charge, and therefore is expected to adsorb important negatively charged solutes such as carbonate and phosphate. Furthermore, when hydrogen sulfide reacts with iron in solution to produce iron sulfide, the reaction can donate electrons to the bound compounds and thereby drive a series of energetically uphill chemical reactions that otherwise could not occur in solution. Wächtershäuser sees these reactions as the beginning of metabolism, occurring on a flat mineral surface rather than in the volume of a cell. He refers to this stage of life's history as the "Iron-Sulfur World." After metabolic processes were initiated in this way, Wächtershäuser proposes that the reaction pathways would somehow become encapsulated in membranes to produce the more familiar forms of cellular life.

The logic and novelty of Wächtershäuser's intellectual construct were impressive. However, as an experimental scientist, I know that an elaborate framework of linked ideas, no matter how elegantly logical, can fall to pieces when a single critical experiment fails. Therefore I was content to await anyone who could falsify Wächtershäuser's hypothesis.

This brings us to the next simulation. Wächtershäuser was a lawyer, not a laboratory scientist, so he needed to find someone with the time and interest to test his ideas. Claudia Huber, an organic chemist at the Technical University in Munich, was just such a colleague, and together they assembled a working simulation of the kinds of chemical reactions that might occur in geothermal conditions such as the chimneys of the deep sea hydrothermal vents described in Chapter 2, in which hot gases are exposed to sulfide minerals. The idea was to heat a dispersion of iron and nickel sulfides together with a source of carbon and see whether anything interesting happened. Something did happen. There was no evidence of a long string of integrated reactions, which would have been the best possible outcome in support of the iron-sulfur concept. But at least carbon-carbon single bonds were produced and acetic acid was synthesized, along with a sulfur-containing form of acetic acid called a thioester. Huber and Wächtershäuser's 1997 paper in *Science* concluded that "the reaction can be considered as the primordial initiation reaction for a chemoautotrophic origin of life."

Huber and Wächtershäuser later published two follow-up papers in *Science*, both related to conditions in which amino acids could be linked together through peptide bonds. Again, they simulated a hydrothermal environment by producing iron and nickel sulfides in the presence of carbon monoxide in boiling water. If amino acids were also added to the mix, they found that these conditions chemically activated the amino acids, which then went on to form peptide bonds. In their 2003 paper, Huber and Wächtershäuser concluded that "the results support the theory of a chemoautotrophic origin of life with a CO-driven, (Fe,Ni)S-dependent primordial metabolism."

So, how do we summarize all these ideas and results? To his credit, Wächtershäuser certainly stirred up the field and challenged other workers to be more critical about their assumptions and open to alternative explanations. And he was courageous enough to follow his own advice, with three papers published in *Science* to prove it. On the other hand, the results are ultimately disappointing in another sense. They represent a way to make a chemical system a little more complex, but so have a number of other simulation experiments. Nor is there any indication that a series of linked reactions can occur on pyrite surfaces. But like all good hypotheses, there are other tests to explore, and the next few years will show whether or not the "Iron Sulfur World" has significant explanatory power.

I agree that a volcanic setting, such as the Kamchatka field site I described in Chapter 2, is a plausible site for the chemical and physical events leading up to the origin of cellular life. But instead of the interface between water and mineral surfaces, it seems more likely to me that the fluctuating conditions prevailing at the edge of hydrothermal springs will be a more fruitful site to model experimentally. Only under those particular conditions do we have a combination of concentrating effects, mineral surfaces that can act as organizing agents, and free energy available to drive the reactions required to assemble the first protocells.

I first heard Harold Morowitz speak at a Gordon Conference, and I will never forget the clarity and persuasiveness of the ideas he presented regarding how phosphate became embroiled in metabolism, the fire of life. We have since published papers together and sailed around his beloved Hawaiian Islands in the small yacht that Harold and his wife Lucille moored in Lihue Harbor on Maui. Harold has spent his life thinking about the patterns of chemicals found in metabolism and bioenergetics, and in a later chapter, I will describe how the core reactions of metabolism might be a kind of fossilized remnant of chemistry that could occur in the prebiotic environment.

Harold spends much of his time as a professor at George Mason University in Virginia, and his ideas attracted the attention of Bob Hazen and George Cody, who are scientists at the Carnegie Institution of Washington, just across the Potomac River in Washington, D.C. A biochemical compound that is central to the metabolic pathways of all life is a three-carbon compound called pyruvic acid, which looks like this: $CH_3–CO–COOH$. As you are reading this, glucose delivered to your brain cells is being broken down into pyruvic acid, which then enters the thousands of mitochondria in every cell to be oxidized to carbon dioxide. This metabolic process releases energy, much of which is trapped in ATP, the energy currency of life.

Harold reasoned that modern metabolism, in which pyruvic acid plays such a central role, is an elaboration of a fundamental series of chemical reactions that would initiate the evolutionary pathway to the intricate network of reactions catalyzed by enzymes today. If so, it seemed reasonable to look for a plausible source of pyruvic acid in a simulated prebiotic environment, just as amino acids, nucleic acid bases, carbohydrates, and lipids had already been added by earlier workers. George Cody, Bob Hazen, and other scientists at the Carnegie rose to the challenge with a geothermal simulation. The idea is to simulate conditions of pressure and temperature found in hydrothermal vents, where the carbon monoxide flows through a hot, iron sulfide mineral at a pressure several hundred times higher than that of our atmosphere. The Carnegie group found that under these conditions, carbon monoxide was adsorbed to the mineral surface and reacted to produce a large suite of organic compounds. One of these, as predicted by Harold Morowitz, was pyruvic acid. In their 2000 paper published in *Science*, the authors concluded that "the natural synthesis of such compounds is anticipated in present-day and ancient environments wherever reduced hydrothermal fluids pass through iron sulfide-containing crust. . . . These compounds could have provided prebiotic Earth with critical biochemical functionality."

To summarize, lots of interesting organic compounds can be produced under simulated geothermal conditions, including pyruvic acid. Bob Hazen and I later collaborated to see what happens when some of these compounds interact with water, and our results will be described in Chapter 7's discussion of self-assembly.

We know from astronomers like Bill Irvine and Lou Allamandola that there is a lot of organic carbon in our galaxy. We also know that hundreds, even thousands, of complex carbon compounds are present in carbonaceous meteorites. How could all of this organic material be synthesized in the cold vacuum of interstellar space, and how did it get into meteorites? That question was central to studies being carried out at the NASA Ames Research Center in Mountain View, California, which brings us to the last simulation in this chapter. Lou Allamandola was inspired by the pioneering work of Mayo Greenberg who, in the 1960s, became convinced that interstellar dust was more than just an obscuring haze that frustrated astronomers as they attempted to visualize faraway stars. Dust, in fact, could provide a place for water, ammonia, carbon dioxide, methanol, and other gaseous compounds to come together in one place, rather than endlessly roaming as isolated molecules in interstellar space. Furthermore, once the gases formed a thin film on the dust particles, the concentrated chemicals could be acted upon by photons of high-energy ultraviolet light, and the activated molecules then had the potential to react and form more complex molecules. Finally, when the dust gathered into a solar nebula during the early stages of solar system formation, they carried the carbon compounds with them, which would explain how complex organics become components of meteorites and comets.

But how can we test this hypothesis? That's where a simulation became essential. Lou Allamandola worked with Scott Sandford, Max Bernstein, and other talented young scientists at NASA Ames to build a simulation of dust grains being acted upon by ultraviolet light. The group could not actually use dust particles, which are microscopic in size, so Lou decided to use a smooth, metal surface cooled to the temperature of interstellar space by liquid helium. The metal is in a vacuum chamber, and small amounts of water vapor, ammonia, and methanol are injected into the chamber where they condense on the ultracold, metallic surface to form a thin film. A high-intensity UV laser light is directed into the center of the metal surface with its frozen film. Then the scientists wait for days, sometimes weeks, condensing several million years of time in a molecular cloud into a fraction of a human lifetime. Finally, the light is turned off, the vacuum is released, and the metal surface slowly returns from the utter chill of outer space to the relative warmth of Mountain View, California, increasing more than 300°C in just an hour or so.

At the start of the experiment, there was nothing in the chamber except three simple gases. If no reactions occurred, the gases would evaporate along with the film of ice, and the metal surface would be just as clean as when it started. But something did happen. Instead of a shiny surface, the metal was coated with a thin yellowish film. In his earlier experiments, Mayo Greenberg had called the product "yellow stuff," perhaps revealing the fact that he was an astronomer, not an organic chemist. Lou, Max, and Scott analyzed their yellow stuff and found all sorts of interesting complex molecules, including the amino acid glycine in one of their more recent experiments.

In 1990, Lou showed me some of the product on the metal surface. Under the microscope, I could see tiny droplets that reminded me of the droplets of "yellow stuff" I extracted from the Murchison meteorite. I asked Lou to do a run, but to use an organic solvent, chloroform, instead of water to rinse the surface. A few days later, I had a small sample of material to work with.

Lou and his coworkers had already analyzed compounds in the film and established that quite a few water-soluble compounds are synthesized. As a membrane biophysicist, my first question was whether there were any components that were *not* soluble in water, which is why I suggested using chloroform as a solvent. The easiest way to check this was to dry some of the chloroform extract on a glass slide and add a dilute solution of slightly alkaline buffer, just as I had with the Murchison extracts. If the products were water soluble, they would simply dissolve, but when I looked at the material under the microscope, the results were astonishing. Not only were there water-insoluble products, but they self-assembled into obvious vesicular compartments, just like the Murchison extracts! This meant that not only had the UV photochemistry produced complex products from a few simple gases, but some of them had the properties of hydrocarbon derivatives that allowed them to form cell-like boundary membranes.

I showed the micrographs to Lou, but at the time, there was no context that allowed us to attach any significance to the observations, so it was just a curiosity. But then, a few years later, astrobiology appeared on the scene as a program supported by NASA, and the link became obvious: Life requires membranes; meteorites contain compounds that can form membranes and deliver them to Earth; and the parent bodies of meteorites in the asteroid belt are composed of dust with organic components produced by UV photochemistry in the molecular cloud. Suddenly, we had a narrative that made sense.

In 1998, Jason Dworkin completed his PhD at the University of California, San Diego, under the guidance of Stanley Miller, and decided to come to NASA Ames to carry out his post-doctoral research. What a rare opportunity! We had Lou, Scott, and Max, all astronomers with an interest in interstellar organics; we had Jason, fully versed in prebiotic chemistry; and we had my own background in membrane biophysics to add to the mix. Jason was the lead member of the team, producing and analyzing tiny amounts of organic compounds in the simulator that had been assembled by Lou and Scott, and I was able to observe the self-assembly process by microscopy. The result was a paper in the *Proceedings of the National Academy of Sciences* with Jason as first author. One of our main points was that carbon anywhere, even in interstellar space, can form compounds that are relevant to life on Earth (Figure 8).

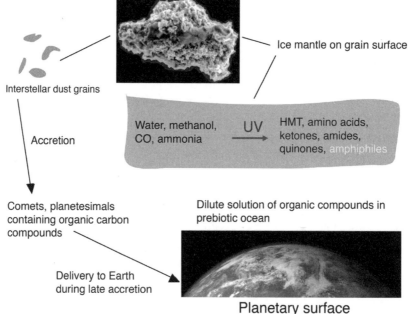

Interstellar dust grains

Ice mantle on grain surface

Accretion

Water, methanol, CO, ammonia

UV

HMT, amino acids, ketones, amides, quinones, amphiphiles

Comets, planetesimals containing organic carbon compounds

Dilute solution of organic compounds in prebiotic ocean

Delivery to Earth during late accretion

Planetary surface

FIGURE 8

Interstellar molecular clouds contain dust grains composed of silicate minerals. These are coated with a thin layer of ice mixed with other simple molecules such as methanol (CH_3OH), ammonia (NH_3), and carbon dioxide. When the ice is exposed to ultraviolet light from neighboring stars, more complex organic molecules are synthesized by a photochemical reaction. During the early stages of solar system formation, the dust particles aggregated into comets and planetesimals which incorporated the ice and organic material, then delivered it to the surfaces of planets such as Earth.

CONNECTIONS

Were the four major classes of biomolecules required for life synthesized from CHONPS on Earth's surface? Or were they delivered as intact organic compounds during late accretion? Surprisingly, the answer to both questions is a cautious *yes*. Cautious, because we don't yet know the relative amounts accounted for by the two alternative scenarios, but we can now connect what we do know with some degree of confidence to the results described in the first three chapters of this book.

We know that stellar nucleosynthesis produces all elements heavier than hydrogen and helium and that the biogenic elements CHONPS are among the most abundant. After being ejected from exploding stars, the elements accumulate on dust grains composed of silicate minerals, which, in turn, aggregate gravitationally into immense molecular clouds. The dust particles aggregate into a star and planets when solar systems form, carrying the organic compounds with them. We can find the organics in comets and meteorites, which represent samples of primitive material from the earliest history of the solar system.

All of the biogenic elements were delivered to Earth during its accretion from the dust and gas surrounding the new sun, mostly in the form of simple compounds such as water, carbon dioxide, and nitrogen. More complex molecules were also present, including biologically relevant compounds like hydrocarbons and amino acids. When oceans appeared, these became reactants in what amounted to a planet-sized flask containing water, atmospheric gases, and volcanic minerals. The mixture was exposed to a variety of energy sources, including abundant light and heat energy and smaller amounts of electrical energy. There was also chemical energy available in some of the molecules and minerals. Because of the constant exposure to energy fluxes, the mixture never reached equilibrium, but instead the energy drove the synthesis of new compounds so that the mixture became increasingly complex in its composition.

Four tiers of chemical reactions then dominate each stage. In the first tier, elements combine into the simplest reactive compounds:

Hydrogen + oxygen $\{\Rightarrow\}$ water H_2O

Carbon plus hydrogen $\{\Rightarrow\}$ methane CH_4

Carbon + oxygen $\{\Rightarrow\}$ carbon dioxide CO_2, carbon monoxide CO

Carbon + hydrogen + nitrogen $\{\Rightarrow\}$ cyanide HCN

Carbon + hydrogen + oxygen $\{\Rightarrow\}$ formaldehyde $HCHO$

Nitrogen + hydrogen $\{\Rightarrow\}$ ammonia NH_3

Phosphorus + oxygen $\{\Rightarrow\}$ phosphate PO_4

Sulfur + oxygen $\{\Rightarrow\}$ sulfate SO_4, sulfite SO_3

Sulfur + hydrogen $\{\Rightarrow\}$ hydrogen sulfide H_2S

In the second tier, the simple reactants combine to form monomers of life:

CO + hydrogen $\{\Rightarrow\}$ alkanes, fatty acids, fatty alcohols

HCN $\{\Rightarrow\}$ adenine

$HCHO$ $\{\Rightarrow\}$ carbohydrates

HCN + $HCHO$ $\{\Rightarrow\}$ amino acids

In the third tier, physical and chemical properties of the monomers lead to more complex structures such as compartments and polymers:

Amphiphiles $\{\Rightarrow\}$ membranous compartments (self-assembly)

Sugars $\{\Rightarrow\}$ polysaccharides (starch, cellulose)

Amino acids $\{\Rightarrow\}$ peptides, proteins

Purines, pyrimidines + ribose + phosphate $\{\Rightarrow\}$ nucleic acids

The fourth tier, the last step before life begins, is the focus of the most recent research. Our hypothesis is that a variety of polymers can become encapsulated in compartments to produce protocells, a few of which happen to have properties that allow them to capture energy and nutrients and then grow and reproduce by catalyzed polymerization. Under the right set of conditions, this might be so probable that in a week or so, a recognizable form of primitive life will self-assemble and begin to grow in the laboratory. On the other hand, even with ideal conditions, it might be so improbable that the origin of life requires a whole planet—with an ocean, volcanoes, and sunlight—to happen just once in 100 million years. We really don't know yet. The reason I am writing this book is to suggest a way to find the answer.

5

THE HANDEDNESS OF LIFE

The large asymmetry in isovaline and other α-dialkyl amino acids found in altered
CI and CM meteorites suggests that amino acids delivered by asteroids, comets,
and their fragments would have biased the Earth's prebiotic organic inventory with
left-handed molecules before the origin of life.

DANIEL GLAVIN AND JASON DWORKIN, 2009

After completing his post-doctoral research at NASA Ames, Jason Dworkin joined the scientific staff at NASA Goddard, in Maryland, where he continues to make important contributions to our understanding of life's origin. The quote above is from a recent paper by his research group, and serves to introduce the topic of this chapter. In the previous chapter, I described how organic compounds can be synthesized from the biogenic elements, and how we study meteorites to learn about this process. But there is more to the story because some organic molecules exist in two forms that are mirror images of each other, just as your left hand is a mirror image of your right hand. This property is called chirality, from a Greek word having to do with hands. Furthermore, the mirror images of something that is chiral cannot be superimposed. In other words, your left hand won't fit into a glove that fits your right hand. To define the "handedness" of such molecules we use the abbreviations L and D from the Latin words laevus and dexter meaning left and right.

Organic chemists know how to synthesize such compounds, but the product always contains equal mixtures of the right-handed and left-handed molecules. Therefore, it was very surprising when it slowly became apparent that living cells were composed of virtually pure left-handed amino acids and right-handed sugars, a property called homochirality. The process by which life became homochiral is probably a deep clue to the origins of life but, as the saying goes, we are clueless. The origin of homochirality is so mysterious that one of the leading ideas is that meteorites were responsible because the amino acids of carbonaceous meteorites have a startling enhancement toward the left-handed versions, as suggested in the quote that begins this chapter.

There is an easy way to understand why life must be homochiral. Imagine a jigsaw puzzle that has 100 pieces. However, the puzzle is different from most puzzles in that it is solved by arranging the pieces in a line, rather than in two dimensions. The puzzle has 20 different asymmetric designs, and the pieces all fit together perfectly by lining up so that the protrusion of one piece fits into the indentation of its neighboring piece. They are homochiral. But if we turn half of the pieces over so that they are upside down, nothing fits and the puzzle can't be assembled. Getting molecules to fit together in polymers is one reason why life must use chiral compounds that are either all left-handed (amino acids) or all right-handed (sugars). The other reason is that the polymers of life always fold into specific structures in order to carry out their function. If the polymers were composed of randomly mixed D and L molecules, they would be unable to fold into the same structure every time they were synthesized.

In his early research career, Gerald Joyce, now a scientist at the Scripps Research Institute, performed a beautiful experiment that clearly illustrates the jigsaw puzzle analogy. Joyce was working with Leslie Orgel at the Salk Institute, who had established a process by which the nucleotide monomers of RNA can line up on a long strand of RNA acting as a template, then link together to form a second strand. (This reaction will be discussed further in Chapter 10.) Joyce and Orgel wondered what would happen if they used a mixture of monomers containing both D and L-ribose. They tried the experiment, and the result was very clear: New strands of RNA were not synthesized! Half of the molecular jigsaw pieces were flipped from D to L, so they were unable to line up on the template to form the chemical bonds of RNA.

WHAT IS THE MOLECULAR BASIS OF CHIRALITY?

It's not too hard to understand why an amino acid or a sugar can be chiral. From Chapter 4, you know that a carbon atom forms chemical bonds with four other atoms. Methane, for example, is CH_4, and carbon tetrachloride, an organic solvent used in dry cleaning, is CCl_4. Now imagine that we replace one of the hydrogens on methane with a carboxyl group (COOH). The result is acetic acid ($H_3C-COOH$). Just from looking at its structure you can see that it is symmetrical, or non-chiral. Next we will add an amine group to make a molecule called glycine, which is still non-chiral because its mirror images can be superimposed. Finally, if we add a methyl group ($-CH_3$), chirality appears. The resulting molecule is the amino acid alanine, and it is chiral because the central carbon has four different chemical groups attached to it. (See Figure 9.)

When chiral molecules are synthesized by ordinary chemical reactions, the resulting product is always an equal mixture of the left- and right-handed versions. However, at some point early in evolution, perhaps even when life began, living systems began to use only left-handed amino acids and right-handed sugars. This is probably an important hint about how life began, but no one has yet found a way to decipher the clue.

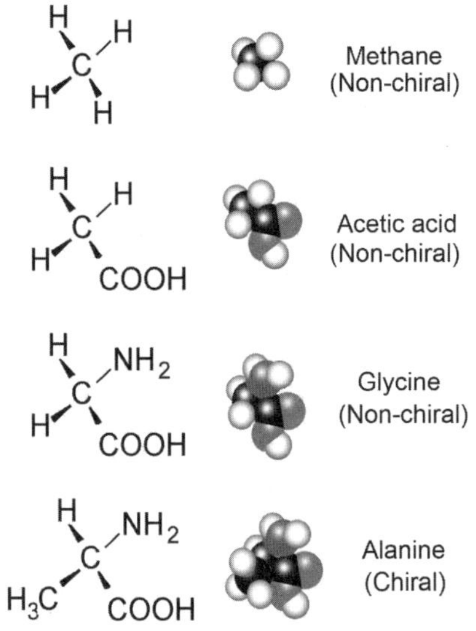

H H
\C/
H—C
H H
Methane
(Non-chiral)

H H
\C/
H—C
COOH
Acetic acid
(Non-chiral)

H NH₂
\C/
H—C
COOH
Glycine
(Non-chiral)

H NH₂
\C/
H₃C—C
COOH
Alanine
(Chiral)

FIGURE 9

Chiral molecules arise when four different chemical groups are bonded to a central carbon atom. This figure shows how addition of chemical groups one by one to a carbon atom can produce the chiral amino acid alanine.

POLARIZED LIGHT AND CHIRAL MOLECULES

The next step in understanding chirality is to realize that ordinary light has some remarkable properties that arise from its wavelike character. The discovery of those properties goes back at least 200 years, when early scientists, including Isaac Newton, began to play with beautiful clear crystals discovered in Iceland. The crystals were perfectly transparent, but when they were laid down on a printed page, instead of seeing the print clearly, as with glass, the lines of print were doubled! Furthermore, if one crystal was placed on top of another, then rotated, each line of print appeared and disappeared in turn. The crystals became known as Iceland spar, and we now know that they are composed of calcite, or calcium carbonate, the same mineral as ordinary limestone but produced by a very slow process so that the calcium and carbonate are arranged into near perfect crystalline arrays.

The effect of calcite on light was very mysterious and fascinated early scientists. Over the next 100 years, it slowly became clear that light had wavelike properties, and it was thought that all of space was filled with something called ether, which transmitted light waves just as air transmits the vibrations called sound. But unlike sound, light waves have three kinds of vibrations, two of which are relevant here. Imagine that you are holding one end of a jump rope, with the other end tied to a post. You can move your hand up and down or sideways so that a series of vertical or horizontal waves are produced in the rope, or you can move your hand in a circle, so that the rope moves in spiral waves. These motions are analogous to light waves, which have properties that are called planar or circular polarization. Planar polarization is most familiar. When light interacts with certain kinds of clear material like the calcite of Iceland spar or the polarized glasses used to view movies in 3D, only light waves with vibrations in a certain plane get through. The main point to be made is that when polarized light passes through a solution containing only one of a pair of chiral molecules, such as L-alanine or D-glucose, the plane of the polarized light is shifted by a certain number of degrees.

Louis Pasteur (1822–1895) was a brilliant French scientist who established that microorganisms like yeast and bacteria were responsible for some of the mysterious changes that occur in stored food and drink. For instance, Pasteur found that heating milk to 56°C kills most of the bacteria that otherwise cause it to sour, a process we still call pasteurization. He also realized that microorganisms were responsible for many of the diseases that afflict humans and domestic animals, and was first to show how immunization by vaccination could prevent diseases like anthrax and rabies.

Winemaking was very important to the French economy in the 1800s. It was common knowledge that during aging, the inside surface of vats became coated with a hard crust. When the compound that formed the crust was isolated and purified, it produced tiny white crystals called tartaric acid, from the word *tartaron* that the Greeks used to describe the same crust 2,000 years earlier. (Dental hygienists also remove hard deposits called tartar from your teeth, but this is caused by bacterial films containing crystals of calcium phosphate, not tartaric acid.) One of Pasteur's first research projects as a young chemist was to crystallize and study tartaric acid. While examining the crystals with his microscope, Pasteur noticed something that no one else had. There were actually two forms of the crystals, and they were mirror images of each other! Pasteur laboriously separated the crystals into two small piles, dissolved them in water and passed polarized light through the solutions. The solution from one of the piles rotated the light clockwise, while the other pile rotated the light counterclockwise. But if the solutions were mixed, there was no effect on polarized light. Pasteur deduced that the tartaric acid had an intrinsic asymmetry to its molecular structure. This was an astonishing stroke of genius, and in 1854, during a lecture where he described how he made such discoveries, Pasteur remarked that ". . . chance favors a prepared mind."

With few exceptions, life uses only amino acids in the L form and sugars in the D form. By the way, another name for tartaric acid is racemic acid, which refers to the mixture of the D and L forms that has no effect on polarized light. The word "racemic" is now used to describe any equal mixture of a chiral compound, and racemization occurs when a chirally pure compound, for instance an L-amino acid, slowly changes over time to a mixture of its D and L forms.

HOW DID HOMOCHIRALITY ARISE?

Chapter 4 described two of the molecules—amino acids and carbohydrates—that are essential to life. Amino acids are polymerized into long protein chains, and simple carbohydrates called ribose and deoxyribose link up with phosphate groups to form the backbone of ribonucleic acid and deoxyribonucleic acid (RNA and DNA). Significantly, these are the two molecular species that underwent chiral selection on the pathway to the origin of life. If the first forms of life used a nucleic acid such as RNA, then a mixture of D and L ribose must have been synthesized by an unknown mechanism and D-ribose was somehow selected when it was incorporated into early nucleic acids or their equivalent. And if peptides were involved in the first forms of life, then L-amino acids were selected. All life is now homochiral, synthesizing only L-amino acids and D-sugars and then using them to construct proteins, nucleic acids, and other polymers involved in living functions.

The prefix "homo-" is derived from a Greek word meaning "same," so homochiral compounds are either all D or all L in their chiral configuration. The origin of homochirality in biology is probably a deep clue to the origin of life, but we have not yet discovered what the clue means. There are, as usual in science, a number of competing ideas, and it is worth describing these in order to illustrate how science goes about testing hypotheses. Here are the three main ideas:

- Homochirality of life is a frozen accident. When life began, the first living cells happened to incorporate L-amino acids and D-sugars and then were stuck with this choice. Early life could equally well have used D-amino acids and L-sugars, but not both.

- An unknown physical process produced slightly more L-amino acids in the environment before life began. This tipped the balance toward homochirality on the L side, but in another solar system, life might have been tipped toward D-amino acids.

- The handedness of life is not an accident. Instead, it is imposed on life by some fundamental force of nature. One possibility is called the weak force, which has a kind of handedness that will be discussed later in this chapter. In other words, homochirality is determined by a law of physics, rather than being a matter of chance. If we discover life on other planets, it will also use L-amino acids and D-sugars.

A FROZEN ACCIDENT?

There is a rule of science called Occam's razor that we try to follow when judging the merit of competing ideas. William of Ockham was a Franciscan monk and philosopher who lived in England in the twelfth century. As a Franciscan, he emulated St. Francis who believed in the philosophy that we would now call "less is more." In other words, a life of simplicity verging on poverty nurtures the spirit. This concept is also referred to as the principle of parsimony, which basically states that given several possible explanations, choose the simplest that accounts for the facts. Parsimony of explanation was a common philosophical belief in Ockham's time, but he wrote so extensively about parsimony that it became associated with his name, now most often spelled Occam. The idea of the razor is that the principle helps to "shave off" unnecessary assumptions.

If we apply Occam's razor to the three possible explanations for homochirality, which explanation is simplest? Most of my colleagues would choose the idea that life became homochiral by chance, so let's examine this by performing what is called a thought experiment. Imagine that life began as a relatively simple system that could use polymers composed of both L- and D-amino acids. Furthermore, there were only six or so amino acids in the prebiotic mix, rather than the 20 that life uses today, and they were synthesized into polymers by a nonbiological process that can produce short peptides up to 12 amino acids long. Let's assume that 12 is just long enough for the peptide to be a catalyst of some essential reaction. Most of the peptides will be mixtures of D- and L-amino acids, but there is one chance in 1,024 (2^{10}) that a given decapeptide will be composed of pure L- or pure D-amino acids. That doesn't sound like much, but in 1 gram of the peptide there would be about 10^{18} homochiral molecules composed of either all D- or all L-amino acids. If a homochiral peptide were able to catalyze an essential reaction more effectively than any of the heterochiral (mixed D and L) peptides, it would be selected during the evolution of primitive life and everything else would be left behind. There was a 50-50 chance that the first such peptide contained all L rather than all D-amino acids. L happened to get there first, and so L-amino acids became the frozen accident.

HOMOCHIRALITY RESULTS FROM A SLIGHT EXCESS

Occam's razor favors the frozen accident if we assume that the prebiotic amino acids were a nearly equal mix of D- and L-forms. But what if we also need to account for a second fact—that they were not equal? This forces us to propose a slightly more elaborate explanation of homochirality because a slight excess of D or L could tip the balance in that direction. Here is the new fact. We know that a variety of amino acids are present in carbonaceous meteorites. Whenever chiral molecules are synthesized by nonbiological processes, the result is a racemic mixture of the D- and L-forms, so we would expect that the meteoritic amino acids would also be racemic. The surprising fact is that they are not. This was first reported in the 1980s by Michael Engel and Bartholomew Nagy at the University of Arizona, who isolated amino acids like alanine from the Murchison meteorite and analyzed them by a technique that could separate D- and L-forms. The results were clear: There was a little more L-alanine than D-alanine. Engel's results were not immediately believed because there was a chance of contamination by traces of L-amino acids from biological sources, but John Cronin and Sandra Pizzarello at Arizona State University made the same observation a few years later when they analyzed several nonbiological amino acids in the Murchison meteorite. This chapter began with a quote from a 2009 paper by Glavin and Dworkin, who determined that the nonprotein amino acid isovaline in two different carbonaceous meteorites, the Murchison and the Orguiel, has a remarkable 15% and 18% enhancement toward the L-form. Most scientists now accept that for unknown reasons there is a slight excess of L-amino acids and that this might have something to do with homochirality in biology.

Well, you might ask, so what? How would this tip the balance toward L-amino acids and D-sugars? Sandra Pizzarello teamed up with Art Weber at the SETI Institute to find an answer. He knew that chiral simple sugars can be synthesized by the reaction of non-chiral precursors and furthermore, that the products, as expected, were racemic. But what would happen if the reaction took place in the presence of a homochiral amino acid? When Pizzarello and Weber did the experiment, they discovered that if an L-amino acid was present even in small excess, the resulting sugar molecule was shifted toward the D-configuration. This was an astonishing result. Somehow the L-amino acid interacted with the components of the reaction and guided the chirality in the direction that life also chose, toward D-sugars. There is still no good explanation for how this occurs, but at least the observation shows how a slight excess can affect the balance.

But how did the amino acids in the meteorite get tipped toward an excess of the L-form? Again, there is no obvious explanation, only speculation. One possibility follows from the fact that circularly polarized light can differentially interact with and destroy one molecule of a chiral pair. Imagine that the amino acids in carbonaceous meteorites were originally synthesized as racemic mixtures on dust grains near a neighboring star. It is known that certain kinds of stars produce polarized light so that the photons impinging on the dust grains would degrade the D-amino acid, leaving an excess of the L-form. When the dust grains accumulated in the asteroids from which meteorites are produced by collisions, the L excess would still be present, and we would discover the excess when we analyzed a meteoritic sample of the asteroid's surface.

If you are a little skeptical of this explanation, your suspicion is shared by knowledgeable scientists as well. Multiple assumptions are required, and there is little experimental evidence that photodegradation could actually work out in space. It is also significant that carbonaceous meteorites contain nearly 100 different kinds of amino acids. Were they all synthesized on interstellar dust grains? It seems unlikely, but this is how science works. An explanation, however tenuous, is given consideration until someone proposes an alternative that better fits the facts. In the case of meteoritic amino acids, we are still waiting for a better explanation.

HOMOCHIRALITY AROSE FROM A CHANCE PURIFICATION PROCESS

Perhaps homochirality did not arise by selection from a racemic mixture, but instead from a process that produced a homochiral product. There are several ways this can occur, one of which follows from Pasteur's original observation. When a racemic compound like an amino acid crystallizes, what usually happens is that both the chiral forms are equally incorporated into the crystal. However, a few compounds crystallize into pure L- and D-forms, such as Pasteur's tartaric acid. It is possible that certain amino acids became sufficiently concentrated to form crystals, for instance by evaporation, and the crystals separated into D- and L-forms. The first life happened to use the pure L-amino acids to synthesize proteins, thus bypassing the problem of synthesizing proteins from a racemic mixture.

Another version of this idea is that a physical process in the prebiotic environment was able to produce homochiral amino acids by adsorption to a mineral interface. For instance, Bob Hazen at the Carnegie Institution of Washington realized that the common form of the mineral calcite (calcium carbonate, or $CaCO_3$) had crystal surfaces with two possible arrangements of the calcium and carbonate. He tested the surfaces for their ability to adsorb the L- and D-forms of aspartic acid and found that the L-form of the amino acid preferentially bound to one of the surfaces, and the D-form to the other. Like crystallization, preferential adsorption to a crystal surface might lead to a slight excess of a D- or L-amino acid in a localized environment, but on a larger scale the enhancement would tend toward zero because the selective calcite surfaces are equally abundant.

Kenso Soai at Tokyo University proposed yet another possible mechanism and supported it with solid experimental evidence. The idea is that many reactions produce pairs of chiral products, but in a few such reactions, the products have the property of promoting their own synthesis, a process called autocatalysis. No reaction, particularly at the start, represents a perfect mixture of D- and L-forms. There will always be fluctuations that result in a slight excess of one of the chiral products. But if the product of that excess is autocatalytic, there is a kind of exponential growth of the product which outruns the other possible chiral product. The result can be a nearly pure L or D product resulting from asymmetric autocatalysis.

Soai and his associates have also shown that certain chiral minerals can tip reactions toward one or the other chiral product. Quartz, which is crystalline silicon dioxide, is one such mineral. The packing of silicon dioxide in quartz crystals has two possible arrangements that are mirror images of each other, which are referred to as d- and l-quartz. (They are also called male and female quartz. You can purchase such crystals as jewelry and wear them as symbols of yin and yang.) Soai discovered that if powdered d- or l-quartz is added to a reaction that normally produces a racemic mixture of products, the microscopic crystals act as seeds that can tip the reaction toward homochirality, with yields ranging well over 90% of one of the products. The significance of such reactions is that they do not simply separate a racemic mixture into homochiral pairs, but instead produce a pure product in solution. Which chiral product is produced is still a matter of chance, with a 50% probability that it will be one or the other. So far, such experiments have been confined to organic chemistry. For instance, one of Soai's experiments involves the reaction of exotic compounds such as pyrimidine-5-carbaldehyde and diisopropylzinc to produce chiral pyrimidyl alkanol. It is important to demonstrate that asymmetric autocatalysis can also produce homochiral products in biologically relevant reactions, so this is an open question for future research.

HOMOCHIRALITY IS DETERMINISTIC

The word deterministic simply means that something is not due to chance, but in fact, must happen according to the action of a physical law. In the first four chapters of this book, I described how the origin of life is intimately related to three primary physical forces that govern everything that happens in the universe. The first is gravity, which gathers hydrogen into stars and heats it to temperatures that initiate the fusion reaction to produce light. The second force is the electromagnetic interaction that governs all of the most familiar properties of matter, including the entire spectrum from radio waves to ordinary light to gamma rays, as well as the chemical reactions that arise from interaction of electrons in atoms and molecules. The third force is the strong force that holds protons and neutrons together in atomic nuclei, but we only have direct experience of its effect when nuclei split to release that energy as radioactivity. There is one more force, however, that we rarely experience directly. This force becomes apparent in the radioactive decay of a neutron. We tend to think of neutrons as being indefinitely stable. For instance, the carbon atoms in our bodies have six protons and six neutrons and, fortunately, the neutrons in the carbon nuclei are stable and don't blow up. But neutrons by themselves are unstable. If we could somehow pop 100 neutrons out of carbon atoms and watch what happens to them, 15 minutes later, half of the neutrons would have undergone radioactive decay into protons and the high-energy electrons we call beta particles. The strong force that holds atomic nuclei together does not account for the instability of isolated neutrons, so neutrons must be held together (until they decay) by the weak force.

One might think that the weak force would have nothing to do with biology, except for one strange fact: The first three forces have a fundamental symmetry in the way they interact with matter, which is called parity. But the weak force is asymmetric, and we refer to this as parity violation. When a radioactive decay process occurs that is governed by the weak force, the particles that are produced have a handedness, and physicists even use the word "chirality" to describe this effect. Furthermore, it is possible to imagine ways that this fundamental chirality might be transmitted to molecular structures.

One of the first scientists to give this serious thought was Abdus Salam, who shared the Nobel Prize in 1979 with Sheldon Glashow and Steven Weinberg for establishing a theory called the electroweak interaction. In 1993, Salam calculated that the energy difference between D- and L-enantiomers of amino acids, given sufficient time, should resolve a racemic mixture of an amino acid to the lower-energy state, or L-enantiomer, in around 10,000 years. Even earlier, in 1966, Y. Yamagata at Kanazawa University had proposed that "the asymmetric appearance of biomolecules is most naturally explained by supposing a slight breakdown of parity in electromagnetic interaction and an accumulation of it in a series of chemical reactions." Other scientists agreed. For instance, Stephen Mason and George Tranter at the University of London calculated the parity-violating energy difference (PVED) for alanine in aqueous solutions, and determined that the energy of L-alanine is lower than D-alanine by $\sim 6.5 \times 10^{-14}$ joules per mole. More recent theoretical calculations of PVED are even higher, in the range of 10^{-12} joules per mole. This difference is extremely small so that amplification of some sort would be required to produce a detectable effect at a macroscopic level.

Most of the scientists working on the origin of homochirality in living systems are deeply suspicious of PVED as an explanation, and rightly so. The calculated energy differences between D- and L-forms of amino acids are so small that they have never been measured. On the other hand, the calculations seem to be correct, so there is a chance that PVED might tip the balance toward L-amino acids if a yet to be discovered mechanism can amplify the difference by many orders of magnitude.

RACEMIZATION IS A PROBLEM FOR THE ORIGIN OF HOMOCHIRALITY

Let's imagine that someone does find a plausible way by which an excess of L-amino acids or D-sugars might be produced on prebiotic Earth. There is still a problem that must be overcome, which concerns racemization. If we have a solution of a pure L-amino acid in the laboratory, and determine that it is absolutely pure, with no detectable amount of its D-form and then make the same measurement a year later, we will discover that a small amount of D is now present. And if the solution is heated to near boiling temperature, some of the D-form can be detected just a day later. One way to understand the effect of heat is to consider 100 pennies arranged all heads up. Every 10 seconds a coin is chosen at random and flipped. There is a 50-50 chance that the coin will land heads or tails, and it is easy to see that over several hours, the coins will approach an equal mixture of heads and tails. If we flip a coin once per second (analogous to heating it up) this will happen more rapidly, and in a few minutes' time, there will an approximately equal mixture of heads and tails.

Like flipping coins, amino acids in solution are continuously switching back and forth between their L- and D-forms at a certain rate, which means that a pure L-amino acid will slowly accumulate its D-form until it approaches an equal mixture of D- and L-amino acids. This process is called racemization, after racemic acid, the D-L mixture of tartaric acid studied by Pasteur. The rate of racemization depends on the temperature and the species of amino acid. At the temperature of ice, racemization occurs very slowly over thousands to millions of years, but at boiling-water temperatures, racemization can occur appreciably in a day or so.

Now consider the origin of life, which likely took place over millions of years. Jeff Bada, at the University of California, San Diego, has made an extensive study of racemization rates, and points out that even if a mechanism is discovered that can produce a solution containing largely L-amino acids, on a geological time scale the homochiral solution will rapidly decay into a racemic mixture. Only if homochirality is produced at a rate fast enough to stay ahead of racemization will life have access to the L-amino acids or D-sugars required to start up metabolism and polymer synthesis. This theory argues in favor of the idea that early life did not require a source of homochiral compounds. Instead, once it got started by selecting the required L-amino acids from a racemic mixture, the selection process simply continued. Microbial life today uses the same process.

Bacteria easily cope with racemic amino acids in a nutrient growth media by using only the L forms and leaving the D-amino acids behind.

CONNECTIONS

The connection between chirality and life processes must be taken into account if we are to understand how life began. All life today is homochiral, and uses only L-amino acids and D-sugars because homochiral molecules fit together into polymers, much as puzzle pieces fit into a jigsaw puzzle. If all the pieces are face up, the puzzle can be assembled, but if half the pieces are turned over (racemized), assembly becomes impossible.

When chiral molecules are synthesized by ordinary chemical reactions, they are racemic, with equal amounts of D- and L-forms in the product. The problem for the origin of life is that the chemical properties of the D- and L-forms are virtually identical, which makes it very difficult to separate them. This is why we still don't know how life became homochiral. One possibility is that the earliest forms of life used racemic molecules and then slowly evolved a mechanism that allowed choice of L-amino acids and D-sugars to produce polymers that were much more efficient in functions such as catalysis and replication. In this scenario, the choice of L and D was due to chance, and life could have equally well used D-amino acids and L-sugars.

There may be more to the story because of the connection to the organic components of carbonaceous meteorites. The Murchison meteorite contains amino acids presumably produced by a chemical process, so they should be racemic. However, there is a small percent more of the L-form, which means that there exists a nonbiological process by which the L-amino acid content of a racemic mixture can be enhanced. It is possible that such an enhancement determined the chiral choice of life on Earth, but there is still much more to learn before we can be confident that we understand the handedness observed in all forms of life today.

ENERGY AND LIFE'S ORIGINS

*For those who want some proof that physicists are human, the proof
is in the idiocy of the different units which they use for measuring energy.*

RICHARD FEYNMAN, 1967

Richard Feynman was a truly brilliant intellect and a remarkable human being. He established physical theory that still guides our understanding of what happens at the subatomic level of physics, for which he won the Nobel Prize in 1965, but he also joyfully played bongo drums in Mardi Gras parades in Rio de Janeiro. His lectures at CalTech are wonderful examples of how to teach with clarity, and the quote above is taken from one of his discussions. The fact is that we spend our lives in a world driven by energy, but it is not a simple matter to define what the word really means, even for physicists. Even harder is to actually measure some form of energy. It is easy to measure a pint of beer (volume), a pound of hamburger (weight), a 200-yard drive (length), or even a minute of time. But what is intuitive about a joule of energy? Or an electron volt? The only energy unit in common use is the Calorie that measures the energy content of food, but most of us would be hard put to recall that a Calorie is the heat energy required to warm a kilogram of water from 14.5°C to 15.5°C. And yet, when we do grasp the basic concepts of energy, we have a much greater appreciation of how the biosphere really works and of ourselves as living organisms embedded in the biosphere.

The best way to gain an intuitive understanding of energy is to discover it by personal experience, just as early chemists and physicists did, so throughout this chapter I will give examples from everyday life. In my high school physics class, energy was defined as "the ability to do work" which sounded pretty tautological because I had the impression that work and energy meant more or less the same thing. But when I finally understood that work is a measurement of change over time, and energy is what causes the change, it made a little more sense. If something changes over time, work has been done and a certain amount of energy has been expended. An important corollary is that energy never disappears. Instead, it is transformed from one kind of energy to another, making something happen along the way, and most often ending up as heat.

The first thing to understand about energy in relation to the origin of life is that mixtures of chemical compounds can contain energy stored in the electronic structure of their bonds. An example is the mixture of hydrocarbons (gasoline) and oxygen (air) in the cylinders of an automobile engine. When an electrical spark ignites the mixture, the oxygen combines with the hydrocarbons to produce water (H_2O), carbon dioxide (CO_2), and heat. The heat causes the gas mixture to expand, which produces a large explosive change in pressure that drives the piston downward.

But energy can also be added to certain organic compounds, which are then able to undergo a chemical reaction that otherwise would not occur. For instance, when chlorophyll molecules in green plants absorb light energy, the light causes the electronic structure to change in such a way that the light energy is captured. The excess energy is released when the electrons pop off the chlorophyll molecules and travel down a complicated chain of reactions to end up on carbon dioxide. As a result, the carbon dioxide is "fixed" in the form of carbohydrates, which are then used as an energy source by other organisms, including human beings.

Some changes produced by energy occur as single events. For instance, a lightning bolt releases a surge of electrical energy that can cause chemical changes in the atmosphere, as Stanley Miller's experiment showed. Other changes are produced by cycles of energy, such as the tides resulting from the rotation of Earth interacting with the moon's gravitational effect on the oceans. Most of the functions of life also occur as cyclic processes, and in this chapter on bioenergetics, the first thing to consider is the surprising fact that phosphate plays a central role in such cycles.

WHY PHOSPHATE?

Phosphate is not an organic compound, but instead is simply a phosphorus atom with four oxygen atoms attached to it, abbreviated PO_4. Over the past 50 years, it has become apparent that life as we know it could not exist in the absence of phosphate, which is ubiquitous in the metabolic processes that drive biological functions. The reason is that the chemical properties of phosphate allow it to be readily linked to organic compounds by enzymatic action, and when that happens the compound becomes chemically activated. In other words, the phosphate adds energy to the molecule, but also provides a kind of chemical handle that allows it to be captured by enzymes so that the molecule can undergo further chemical changes associated with metabolism. The addition of phosphate has a name: phosphorylation. Other inorganic ions, such as sulfate, were much more abundant on prebiotic Earth, but phosphate has just the right set of properties that made it the preferred species for chemical activation.

In virtually every metabolic pathway one or more of the components must be phosphorylated to activate the process. Sometimes the enzymes themselves must be phosphorylated in order to be active, and this is very important in signaling and regulatory pathways. For instance, muscles are caused to contract by the release of calcium ions from membranous storage sites within cells, called sarcoplasmic reticulum, and the calcium must be pumped back into the storage sites if the muscle is to relax. The pump is an enzyme called calcium ATPase, which is embedded in the membranes and undergoes a cyclic phosphorylation by ATP as it transports calcium ions back into the storage sites. The remarkable biochemical pathway that underlies vision also involves phosphate. As light passes through the lens to form an image on the retina, individual photons are absorbed by an eye pigment, called rhodopsin, in the membranes of rod and cone cells. The rhodopsin activates a protein called transducin, which amplifies the signal because 100 transducins are triggered by each rhodopsin. The transducin in turn activates an enzyme called phosphodiesterase, which amplifies the signal further by hydrolyzing 1,000 nucleotides called cyclic guanosine monophosphate (cGMP). There is more to this story, but the ultimate result is that an electrical signal is generated that travels to the visual cortex in the brain, which integrates all the incoming signals into the image of the page that you are reading.

An important, still unanswered question concerns how phosphate might have first become involved in life processes. The problem is that phosphate on Earth's surface is rare in solution because it is mostly present as a mineral called apatite, the same combination of calcium and phosphate that composes tooth enamel and bones. From the fact that teeth don't dissolve during a human lifetime, it is obvious that apatite has a very low solubility. So where did the phosphate come from that now plays such an important role in metabolism? There are no convincing explanations yet. If I had to guess, it would be that life began near a mineral deposit of apatite that was exposed to a low pH similar to that of volcanic puddles. This is an acidic pH range in which calcium phosphate can dissolve and release phosphate anions. The presence of free phosphate in solution permitted its initial incorporation into organic compounds as phosphate esters. A second series of chemical reactions could then be initiated, leading to simple metabolic pathways involving phosphate.

ATP IS THE ENERGY CURRENCY OF LIFE

Now let's talk about the intracellular phosphate cycle. Phosphate in the cytoplasm diffuses to mitochondrial membranes where energy is used to attach it to adenosine diphosphate (ADP) to make adenosine triphosphate (ATP). The energy is stored in the chemical bond between the second and third phosphate on ATP. The ATP then diffuses throughout the cytoplasm where the stored energy is released to perform some cellular function such as activating metabolic reactions, driving ion transport across membranes, or causing a muscle cell to contract. When the energy in ATP is used, phosphate is released from ATP, and the resulting ADP and phosphate then diffuse back to the mitochondria to repeat the cycle.

ATP is called an energy currency because, like money, it does not stay in one place, but instead diffuses around and makes things happen. The earliest life needed some form of energy currency, perhaps something like ATP, so we need to understand the structure of ATP and how it can transport and release chemical energy. An ATP molecule is shown in Figure 10. It is surprisingly simple, just an adenine, a ribose, and three phosphate groups linked through chemical bonds. As noted in Chapter 4, adenine can be produced in prebiotic simulations when five HCN molecules polymerize. Adenine is also one of the many organic compounds present in carbonaceous meteorites, so it is likely that small amounts were present in the prebiotic environment. The sugar ribose is one of the carbohydrates that form when formaldehyde reacts with itself, and simple carbohydrates have also been found in meteorites. But even if they were present, how did adenine, ribose, and phosphate get together to produce ATP?

This is an open question. Unlike the facile reactions that produce adenine and ribose from HCN (hydrogen cyanide) and HCHO (formaldehyde), there are no obvious pathways to ATP, or even to adenosine monophosphate (AMP) and other nucleotides that are the monomers of RNA and DNA. But maybe ATP was not required as an energy source for the first living cells. Is there something simpler?

In fact, there is. The chemical energy of ATP is contained in the anhydride bonds that link the last two phosphates of the triphosphate chain. Anhydride bonds are synthesized when two acidic groups lose a water molecule and form a covalent bond. The chemical formula for water is usually written as H_2O, but here I will write it as H–OH to keep

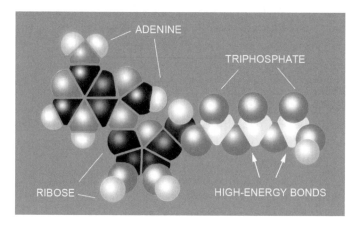

FIGURE 10

Adenosine triphosphate (ATP) is the primary energy currency of all life. It is composed of adenine, ribose, and three phosphate groups, the last two linked through pyrophosphate bonds. The chemical energy is contained in the pyrophosphate bonds, and the energy is used by numerous enzymes to activate other molecules in the cell by adding phosphate to their molecular structure.

track of the hydroxyl group (–OH). Anhydride bonds are always high in energy content and have the potential to release that energy when water is added back to the bond (hydrolysis by H–OH) or by transferring one of the groups to another molecule, usually to a hydroxyl group (–OH) on an organic compound (R–OH).

It turns out that phosphate has the unique ability to form anhydride bonds with itself in a very simple reaction:

$$\text{Phosphate} + \text{phosphate} + \text{heat} \{\Rightarrow\} \text{pyrophosphate} + H_2O$$

Pyrophosphate gets its name from *pyro*, the Greek word for "fire," because it is easily synthesized by heating phosphate under dry (anhydrous) conditions. The reaction can continue to add phosphate groups so that long chains called polyphosphate are synthesized. It is known that pyrophosphate is produced when phosphate is exposed to dry heat in volcanic conditions. Just as apatite is a common mineral composed of calcium and phosphate, another relatively rare mineral called wooldridgeite has been discovered that contains pyrophosphate. Furthermore, microorganisms such as yeast and certain bacteria use pyrophosphate and polyphosphate to store energy. The bottom line is that we are not restricted to ATP as the only possible energy currency for primitive life. Pyrophosphate is a much simpler molecule with a similar energy content, and therefore, it is a viable alternative.

WHAT SOURCES OF ENERGY WERE AVAILABLE ON EARLY EARTH?

Sometimes energy is invisible, like the energy content of food or gasoline, but we can directly sense other forms of energy like wind, tides, and sunlight. Furthermore, some of the energy available on early Earth caused changes in a single event, such as a lightning bolt or a comet impact, while other forms were available continuously as cycles. The main point to remember is that energy is defined by its effect on matter, which is to produce change over time. We can use this definition to think about the kinds of energy that were available on early Earth to drive chemical reactions. In other words, what was changing over time? Because all of chemistry boils down to changes in the electronic structure of atoms and molecules, what we are really asking is what forms of energy could cause the electronic structure of organic compounds to change over time. There are just a few energy sources that we need to consider:

- Energy is contained in the paired electron bonds that hold atoms together in molecules. This is called chemical energy.
- Energy is made available when electrons jump from one substance to another. This is called electrochemical energy, sometimes referred to as oxidation-reduction energy or redox potential.
- Photochemical energy is added to chemical compounds when light is absorbed by the electronic structure of a pigment molecule.

- Heat energy is added to compounds when the temperature increases.
- Energy is available in concentration gradients across membranes. This is chemiosmotic energy, which is the key intermediate between light energy and the rest of life on Earth.

Living organisms today use four of the five sources of energy to drive a variety of processes that are essential for life to exist. (Heat is the exception.) After a source of energy available in the environment is captured, fairly complex enzyme-catalyzed reactions transform that energy into other forms of energy before it can be used to drive cell functions, a process called energy transduction. In the rest of this chapter, I will discuss each of the energy sources in turn and show how they are used by contemporary organisms. Then we can make some educated guesses about how energy sources on prebiotic Earth could drive organic compounds toward ever-increasing complexity until self-assembled systems of molecules gave rise to the origin of cellular life.

CHEMICAL ENERGY

Two fundamental concepts are used to describe how energy drives chemical reactions, and it is essential to grasp these in order to understand how chemical energy would be involved in the origin of life. The two concepts are generally classified as *thermodynamics* and *kinetics*. Thermodynamics is used to describe the energy changes that occur during a chemical reaction, while kinetics describes the rates at which a given reaction occurs.

Although these terms sound abstract at first, chemical reactions begin to make a lot more sense once you get used to thinking in terms of thermodynamics and kinetics. The easiest way to understand these two concepts is by analogy. Imagine that we have a dozen marbles in a pan and that a small motor lifts the marbles one at a time to a height of 1 meter and then releases them into a pan at the top that serves as a reservoir. The motor uses energy to lift the marbles up, and that energy is stored in the marbles at the top. The marbles can then take two possible paths to get back down. One path leads directly back to the lower pan simply by falling, but the second is a longer track with many curves and loops. (Some airports have versions of this device to entertain passengers waiting to board their plane.) A marble that falls back to the pan releases its energy all at once, but a marble taking the second pathway releases the energy much more slowly and can perform all kinds of interesting tricks on the way down. The point is that exactly the same amount of energy is used to raise the marbles up (thermodynamics) but the stored energy can be released quickly or slowly (kinetics).

THE ENERGY OF CHEMICAL REACTIONS

The primary thermodynamic concepts that describe chemical reactions are free energy, enthalpy and entropy of a reaction, equilibrium and disequilibrium, and activation energy. Let's talk about equilibrium first. Most of us have played with baking soda and vinegar, perhaps mixing them together in a balloon and watching the balloon expand. The active chemicals (the reactants) in these common products are sodium bicarbonate in the baking soda and acetic acid in the vinegar. A chemist would write the reaction as the following equation:

$$NaHCO_3 + CH_3COOH \{\Rightarrow\} CH_3COONa + H_2CO_3$$

In other words, the acidic hydrogen ion on the acetic acid replaces the sodium ion in the sodium bicarbonate to make sodium acetate (CH_3COONa) and carbonic acid (H_2CO_3). Then a second reaction occurs in which the carbonic acid breaks up into water and carbon dioxide:

$$H_2CO_3 \{\Rightarrow\} H_2O + CO_2$$

When the reactants are mixed, a chemical reaction occurs with lots of fizzing due to the carbon dioxide that is produced, which also blows up the balloon. But after a minute or so the expansion slows and finally stops. The reaction has reached a state called equilibrium. The most important point to understand about equilibrium is that the reaction does not actually stop. The forward reaction continues at a certain rate, but it is increasingly balanced by the reverse reaction in which carbon dioxide reacts with water to produce carbonic acid. When the two rates are equal, the reaction has reached equilibrium, so we should then write the equation with arrows going in both directions:

$$NaHCO_3 + CH_3COOH \{\Leftrightarrow\} CH_3COONa + H_2CO_3 \{\Leftrightarrow\} H_2O + CO_2$$

The concepts of equilibrium and disequilibrium are central to the metabolic reactions of life today, as well as the reactions that led to the origin of life. To summarize:

- Every chemical reaction has a forward and reverse direction, and every reaction is, at least in principle, reversible.
- Chemical reactions always proceed from a state of disequilibrium to a state of equilibrium.
- The change in energy content that occurs as a reaction proceeds toward equilibrium is called the free energy related to that reaction.

Biochemists often describe the free energy of a reaction in terms of kilocalories per mole, but this unit is slowly being replaced by the standard international unit of kilojoules per mole. The joule was named to honor James Prescott Joule, a British physicist who discovered in 1845 that mechanical energy could be turned into heat energy. Joule observed that if a certain amount of mechanical energy is used to stir water, the water gets warmer by an equivalent amount of heat energy. The modern definition of a joule is the energy added to 1 kilogram of mass when 1 newton of force acts on it for 1 second. But let's make this less abstract. Imagine that you are an astronaut in the International Space Station, and you have a 1-liter bottle of water that you want to send to another astronaut in the same room. If you push on the bottle for 1 second so that it ends up floating across at a velocity of 1 meter per second, you used 1 joule of energy to make it happen.

The next thing to understand is that when we delve deeper into the free energy changes during a reaction, we discover that free energy has two components called enthalpy and entropy. Furthermore, a chemical reaction always requires a special kind of energy called activation energy to get underway, so we need to know about this to understand what enzymes and catalysis are all about. Let's first talk about free energy, enthalpy, and entropy.

Free energy is the total energy change that occurs during any reaction. It is called "free" because the compounds that are reacting contain much more total energy than is actually used during the reaction, so the free energy refers only to the energy that is "freed" or expended when a reaction occurs. The free energy change is negative when a reaction is spontaneous, which you can remember by recalling that energy is lost from or subtracted from the reactants. When energy must be added to make a reaction occur, to drive it "uphill," the free energy change is positive.

Most spontaneous reactions give off heat, so early chemists tried to understand the energy of a reaction in terms of heat. This is where the term "thermodynamics" comes from. "Enthalpy" is derived from Greek words having the sense of "internal heat," and enthalpy changes can be easily measured, because most spontaneous reactions release energy as heat, such as the explosive reactions in a car engine or the hydrolysis of ATP that keeps your body warm. (This is why you get hotter when you run—because of the extra ATP hydrolysis that is required to contract and relax leg muscles.) However, different reactions have different enthalpy changes. For instance, if we burn 1 gram of fat and 1 gram of sugar and compare the amount of heat released, we find that fat releases twice as much heat as sugar. The reason is simple. Fat is mostly hydrocarbon, with each carbon having two hydrogen atoms attached, and the chemical bond between the hydrogen and carbon is where most of the chemical energy resides. Sugar is a carbohydrate that is already partially oxidized, with one hydrogen and one hydroxyl group (–OH) on each carbon, and therefore contains less chemical energy. The products of combustion, H_2O and CO_2, have lost all of their chemical free energy.

Entropy was discovered in the late 1800s, when chemists tried to describe the energy of chemical reactions entirely in terms of the heat given off. It just didn't add up. Sometimes the heat energy released was almost (but not quite) what was expected, but in other reactions it was not even close. There was something missing. In 1865, German physicist Rudolf Clausius gave it a name: "entropy," derived from Greek words meaning "transformation within." A few years later, Josiah Willard Gibbs at Yale University came up with the concept of free energy, meaning the energy available to do external work, which is now abbreviated G, or Gibbs free energy. Enthalpy is relatively easy to grasp, but entropy has a mysterious quality when first encountered. Gibbs wrote in 1873 that "any method involving the notion of entropy, the very existence of which depends on the second law of thermodynamics, will doubtless seem to many far-fetched, and may repel beginners as obscure and difficult of comprehension." Gibbs realized that the missing energy we call entropy is not available to do external work because it is used up internally in a reaction as the reactants change into products. Therefore, it must be subtracted from the available energy according to the following simple equation:

G (Gibbs free energy) = H (heat or enthalpy) − TS (absolute temperature times the entropy)

In the laboratory there is no need to measure the total energies of the reactants and products of a reaction, which in fact would be very difficult. What is of interest and much easier to measure is the *change* in energy as the reaction occurs. Therefore, the most useful form of the above equation has an additional abbreviation Δ, meaning difference or change:

$$\Delta G = \Delta H - T\Delta S$$

ΔH is easily measured by the amount of heat energy given off during a reaction, which might be determined, for instance, by the change in temperature if the reaction occurs in water. Entropy, abbreviated S, cannot be directly measured but must be calculated from other measurements. An intuitive way to understand entropy is from our general experience that, left to themselves in a closed system, orderly things tend to become disordered over time. The increase in disorder is defined as an increase in entropy. During most chemical reactions, entropy increases, and the greater the change in entropy, the larger the TΔS term becomes. In a few reactions, there is almost no change in enthalpy, and the reaction is driven primarily by the change in entropy. Important examples of such processes in biology include the self-assembly of lipids into bilayer membranes and the folding of protein into tertiary structures.

T is the absolute temperature in degrees Kelvin, which also needs a little explaining. Early chemists decided to choose the melting and boiling points of water to define temperature measurements, and this historical accident has caused endless confusion. In 1848, an Irish physicist named William Thomson published his idea that there must be a minimum temperature below which nothing could get colder, and he defined this temperature as absolute zero. Thomson arbitrarily chose the familiar Celsius scale to define a degree, in which there are 100 degrees between the melting and freezing points of water. Later in his career, Thomson was knighted for his scientific and engineering achievements and given the title Lord Kelvin after the river that flowed past the University of Glasgow, in Scotland, where he spent most of his career. By this circuitous path, the name of a Scottish river was chosen for the units of absolute temperature, which are called Kelvins, abbreviated K.

For our purposes, all you need to remember from this explanation is that when a reaction can occur spontaneously, the change in free energy is negative by definition because energy is lost from the reactants as heat (the enthalpy change) and because the products have a greater degree of disorder (entropy) than the reactants. For example, consider striking a match. The reactants are the chemicals in the match head, and they are all together in one place. But when the match ignites, heat is given off (the enthalpy change) and the reactants turn into gases which become disordered (increase in entropy) as they diffuse into the surrounding atmosphere.

There is one last point to make about thermodynamics, which has to do with what we call open systems and closed systems. A closed system is just what it sounds like, something going on in a container that has no access to the outside world. We have found no exceptions to the rule that any reaction going on in a closed system ultimately reaches equilibrium in which no further energy changes occur. All of the free energy has been spent, and entropy has reached a maximum. This law of the universe has been used to mystify the origins of life. The argument goes something like this: If thermodynamics says that everything always goes toward disorder (equilibrium and maximum entropy), how could life begin? The answer is simple, of course. Life did not begin in a closed system. Early Earth, like Earth today, was an open system in which an immense amount of impinging solar energy (and stored volcanic energy) was available to drive chemical reactions toward ever-increasing complexity. Solar energy remains the power source for the biosphere and will continue to be for the next several billion years.

KINETICS AND ACTIVATION ENERGY

Some chemical reactions are very slow and others are very fast. There are equations that let us describe how fast reactions occur, but for our purposes, we don't need to go into detail. The most important thing to understand is that some reactions can be driven energetically uphill very fast, but if the downhill reaction in the other direction is slow, it is possible to make complicated molecules and keep them in a metastable state for extended periods of time. We call this condition a kinetic trap. In one sense, that is what life is all about. Life is thermodynamically far from equilibrium, yet life can exist because the downhill reactions of degradation are slow. In other words, we don't dissolve when we take a shower, even though that is a spontaneous reaction in terms of thermodynamics. What this means for the origins of life is that we need to discover a process by which polymers can be synthesized rapidly, but the resulting polymers are kinetically trapped and therefore accumulate long enough to take part in the biochemical functions of life.

But what about activation energy? This is the reason that some reactions are very slow, such as hydrolysis, while others are fast, like the reaction between baking soda and vinegar. The idea of activation energy is that a reaction is never purely a downhill process. In order to undergo a chemical reaction, two reactant molecules must collide as they diffuse in a gas or a liquid. During the collision, there is a kind of energy hill that the reactants must go over before they can proceed energetically downhill to become products. The energy hill is called activation energy, which in some reactions is relatively high, but in others it is very low. Without realizing it, we often experience activation energy in everyday activities. For example, kitchen matches are loaded with chemical energy, but they are inert until you add activation energy by scratching them on a hard surface. The scratching produces heat by friction, and the heat is enough to activate the reaction in a small portion of the match head which then spreads by a chain reaction to the rest of the match.

The point here is that when we add heat to a reaction, we add kinetic energy to the molecules in the mixture. In other words, hotter molecules move faster then cooler molecules so that when they collide, they are more likely to surmount the activation energy barrier. This fact is used by organic chemists who almost always heat a reaction mixture to make the reaction go faster. It is also the reason that you bake bread in an oven to produce desirable reactions in the dough. A second reason to understand activation energy is that reaction rates can be dramatically increased by catalysts even without heating because catalysts lower the activation energy hill faced by reactants. This concept is so important to life, and to the origin of life, that a later chapter in this book is entirely devoted to catalysts.

HEAT AND ACTIVATION ENERGY

Heat is a ubiquitous form of energy that was abundant in the prebiotic environment. For this reason, early researchers investigating the origins of life attempted, with some success, to drive polymerization reactions by drying potential reactants such as amino acids at elevated temperatures. The advantage of using dry heat to drive condensation is that it is a plausible way to synthesize prebiotic polymers. RNA can be synthesized nonenzymatically on a template, as demonstrated in the 1980s by Leslie Orgel, or on the clay surfaces studied by Jim Ferris at Rensselaer Polytechnic Institute. However, these simulations of prebiotic polymerization mechanisms require that monomers are chemically activated in some way, typically by having a "leaving group" attached to the phosphate through an ester bond. This means that water has already been removed from the reactants so that the polymerization reaction is energetically downhill. No heat is required, and the reactions occur in aqueous solutions. These conditions are ideal from a chemist's perspective, but no one has yet thought of a way to chemically activate monomers under prebiotic conditions. This is why dry heat is an interesting alternative approach for producing the primitive polymers required for life to begin.

An important point to be made here is that heat is not a true source of chemical energy. You can heat a solution of amino acids in water forever, but no polymers will be produced. Even if a few peptide bonds form, they are rapidly hydrolyzed by the water. On the other hand, polymer synthesis can occur if the amino acids are in a dry state, so the actual source of energy is the fact that water molecules have the potential to leave the system when two reactive molecules happen to come together. What heat does is to add activation energy to the reactants so that the reaction is more likely to occur. The water essentially evaporates, and as a result the reaction is "pulled" toward the synthesis of chemical linkages such as peptide and ester bonds.

The problem with simple heating and drying is that multiple reactions become possible at elevated temperatures. As a result, many nonspecific chemical bonds form, and this can produce an undesirable brown or even black substance called tar. But what if we can find a way to organize reactants so that they do not form multiple varieties of chemical bonds, but instead just the kinds of bonds associated with the living state? The dry heat would then drive the monomers toward specific kinds of polymers. Furthermore, if the dry portion of the reaction were cycled repeatedly through a wet portion, the reaction would be "pumped" uphill with ever more complex products accumulating in a kinetic trap. In this way, a variety of random polymers could be produced, and if amphiphilic molecules were also present, the result would be vast numbers of microscopic protocells, each different from all the others.

METABOLIC ENERGY

The synthetic reactions related to life processes are not spontaneous. If you added up the free energy of the reactants and products during protein synthesis, for instance, you would discover that the free energy change is positive, meaning that energy must be put into such reactions to make them go energetically uphill. How this happens is the subject of the rest of this chapter.

Living systems use chemical reactions to release energy in small steps we call metabolism, which can be defined as a series of chemical reactions linked in a molecular system that provides energy and small molecules required for growth. Each step is catalyzed by a specific enzyme. Life today can extract chemical energy from a variety of nutrients, but we will focus on glucose, which is at the center of metabolic pathways. The metabolic pathway by which the energy contained in glucose is released in cells is aptly named glycolysis, taken from Greek words meaning "sugar breaking," which is exactly what happens. When glucose enters the cell, it is carried in by a transport protein in the membrane. The first step is phosphorylation by an enzyme called hexokinase, which catalyzes the transfer of a phosphate group from ATP to the glucose. This is called activation and is essential for initiating glycolysis. Glucose by itself is virtually inert, but the addition of phosphate provides a kind of handle by which further catalyzed reactions become possible. Activation by phosphate addition is not confined to glycolysis. It is a very common reaction in metabolism.

This is not a biochemistry text, so I won't discuss the enzymatic steps in detail. The main point to understand is the way that phosphate additions drive the reaction. The high-energy phosphates of two ATPs are used to activate intermediates, but four ATPs are synthesized, so glycolysis is a net source of energy for the cell. The process is called glycolysis because in one of the steps, a six-carbon sugar breaks down into two three-carbon sugars, each of which ends up as pyruvic acid. If oxygen is available, the pyruvic acid enters mitochondria and is completely oxidized to carbon dioxide and water. Much more energy is available in oxidation, which is conserved in the 36 ATP molecules that are synthesized for each glucose molecule that enters the glycolytic pathway.

This is all pretty complicated! And yet, glycolysis is one of the simplest metabolic systems in cells today. The question is: How did it all begin? There is no consensus yet about how metabolic systems may have originated, so this is a prime problem for future research.

ELECTROCHEMICAL ENERGY

In the middle of winter, when I go out to feed Mystie, our golden retriever, it can be pretty dark early in the morning, so I keep a flashlight handy. I click the switch, the bulb glows, and I can find the dog food in our garage. What happens when I click the switch? Are the flashlight cells somehow full of electricity that causes the bulb to glow? The truth is that flashlight cells are full of chemicals, not electricity, and a chemical reaction produces the electrical current that lights the bulb. It's the same in a car battery with six cells, each containing a chemical reaction between lead and sulfuric acid that produces about 2 volts. The six cells are linked together in series, adding up to 12 volts and enough electrical current to start the engine. And it's the same in the mitochondria in the 100 trillion cells of the human body, in which a chemical reaction between oxygen and electrons produces 0.2 volt and enough energy to synthesize ATP. All of these reactions are classified as electrochemistry, and the question is whether such reactions on early Earth provided energy for the first forms of life.

The electrochemical reactions used by most life today require oxygen produced by plants. When we breathe to bring oxygen into contact with the blood stream so that it can be delivered to the rest of the body, we call the process respiration. This same word is used in biochemistry to indicate the process by which mitochondria use oxygen. The reason oxygen is so important is that it is an electron acceptor, providing a kind of sink into which electrons derived from sugars and fats flow. An analogy is a waterfall like Niagara, in which water flows from a higher level to a lower level. The energy of the falling water can be turned into electrical energy by sending the water through a turbine and generator. But now imagine that we raise the lower river level to the point where there is no more waterfall. The generators grind to a halt, and nearby cities in New York go dark. The same thing happens to a human being in the absence of oxygen. There is no place for electrons to go, and our brain goes dark, followed quickly by the rest of our body.

What this means is that the first forms of cellular life needed both a donor of electrons at the top of the waterfall and an acceptor at the bottom. Several potential donors were likely to be available in the prebiotic environment. Perhaps the most plausible is hydrogen gas itself, as well as hydrogen sulfide (H_2S) and methane, all of which are present in gases released by volcanoes. A variety of microorganisms use these gases as a source of electrons, a good example being the abundant bacteria present in hydrothermal vents. But what are some possible electron acceptors? We know that little or no oxygen was present in the early atmosphere, so the electrochemical energy used by most life today was unavailable. What electron acceptors might have served instead of oxygen? We can find clues to alternative electron acceptors in the metabolism of anaerobic bacteria, which even today thrive under conditions where no oxygen is present. Such bacteria use sulfate, nitrate, iron, manganese, and carbon monoxide. When sulfate is used by anaerobic bacteria as an electron acceptor, one of the products is hydrogen sulfide, the rotten egg smell of swamps and tidal flats. (The term "rotten egg smell" comes from the fact that bacteria occasionally infect eggs and produce hydrogen sulfide when they metabolize sulfur-containing amino acids, cysteine and methionine.)

PHOTOCHEMISTRY

The amounts of energy available from various sources on early Earth can be roughly estimated from those on today's Earth. Light energy wins hands down, with 624,000 kilojoules per square meter per year ($m^{-2}yr^{-1}$), and photosynthesis is the energy source of virtually all life today, which has evolved to take advantage of that source's natural abundance. Much smaller energy sources include radioactivity, volcanoes, and electrical discharges, with estimates ranging from approximately 5 to 200 kilojoules $m^{-2}yr^{-1}$. Could light have been the primary energy source for the first forms of life? To understand this question, we need to know a few facts about photochemistry.

The most important fact is that light only becomes a useful energy source if there is a pigment to absorb the photons. When photons are absorbed by a pigment molecule, they are actually interacting with the electronic structure of the bonds holding the molecule together, thereby adding energy to the bonds. When this happens, we speak of the molecule going from a ground state to an excited state. The energy that is absorbed can be given off in any of a number of ways. Often it simply adds to the normal vibrations and rotations of the molecule. In other words, it turns into heat. Another possibility is that it is given off as light at a longer wavelength, a phenomenon called fluorescence. For example, when ultraviolet photons hit a pigment molecule like fluorescein or rhodamine, the absorbed energy is given off as a bright green or red fluorescence. But the most important process related to life occurs when the absorbed light energy causes the pigment to become more reactive. One such reaction occurs when the activated pigment molecule donates one of its electrons to an acceptor molecule. This seemingly simple photochemical reaction is the energetic foundation of most life on Earth because this is what happens when a green chlorophyll molecule absorbs the red light of sunlight. (The only exceptions to this rule are certain bacteria called lithotrophs that use the energy stored in certain minerals.)

Figure 11 shows a simplified diagram of this reaction. Starting with chlorophyll (square with a tail) in its ground state, a photon of red light is absorbed and increases the energy content of chlorophyll. The added energy causes it to go to an excited state, shown with a star. It can then donate an electron that passes through multiple reactions to carbon dioxide, which then ends up after further reactions as a carbohydrate such as glucose. In this way, the original light energy is conserved in the form of chemical energy stored in glucose. After it loses the electron, chlorophyll is positively charged, and the electron is replaced from a water molecule in the "water splitting reaction," which releases oxygen. This is the source of virtually all of the oxygen in Earth's atmosphere.

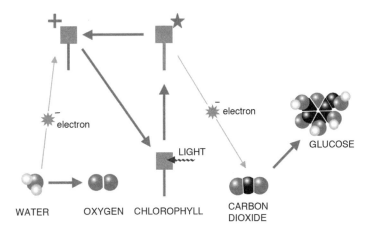

FIGURE 11

In photosynthesis, light energy causes chlorophyll to enter a high-energy excited state that releases an electron. The electron travels through a series of reactions to carbon dioxide, which is reduced to a sugar such as glucose. In a second reaction, the chlorophyll recovers an electron from water with the result that oxygen is released. Photosynthesis is the source of virtually all oxygen in the atmosphere.

If photochemistry is essential for life today, could it also have provided energy for the origin of life? Perhaps, but there is a problem: What pigment molecules were available? Certainly not chlorophyll, which is a very complex molecule requiring multiple enzymatic steps for its synthesis. The best guess is that the first forms of life were heterotrophic and only developed photosynthesis at a later evolutionary stage. Even if we can think of some examples of primitive pigments, the capture of light energy would not be possible unless there were membranes available that could contain the pigment molecules. Furthermore, in order to process the absorbed light energy, a primitive cell would also require a system of other molecules to transduce the light energy into chemical energy, as shown by the arrows in Figure 11. Future research might someday discover a plausible mechanism by which this seemingly complex series of reactions could arise, but until then it seems reasonable to think that some energy source other than light was used by the first living cells.

MEMBRANES, VOLTAGE, AND CHEMIOSMOSIS

Now we come to one of the most abstract sources of biological energy, and yet one that is central to all life today. This energy source is so strange that it took biochemists nearly a century to grasp it after Willard Gibbs came up with the idea of free energy. And it required a maverick scientist named Peter Mitchell 15 years to get his peers in science to accept the idea, which ultimately produced a true paradigm shift in our understanding of how living organisms capture energy.

We have already described how energy can be transformed from one form to another, a process called energy transduction. We have also discussed the most common versions of energy transduction, such as the transformation of one form of chemical energy into another and the transformation of light energy into chemical energy. The transduction of chemical energy is fairly straightforward, so for many years, biochemists thought that the energy source of living cells was simply the energy of chemical reactions called metabolism and that ATP, the energy currency of life, was synthesized chemically.

So imagine the fight Peter Mitchell had on his hands when he proposed the wild idea that ATP was instead synthesized by an amazing chain of coupled processes that had an absolute requirement for the existence of membrane barriers and proton gradients. I'll explain this in a moment, but first let's get an overview of the entire process from start to finish. Mitchell was trained as a microbiologist, and was well aware of the fact that bacterial cells are basically compartments bounded by membranes. Furthermore, the mitochondria and chloroplasts that transduce energy in animal and plant cells are descended from the original symbiosis between bacterial species that gave rise to eukaryotic life. In the late 1950s, it was becoming clear that the membranes of bacteria, mitochondria, and chloroplasts contain systems of proteins and cofactors that are able to transport electrons from one set of compounds to another. We already described how this occurs during photosynthesis, in which the source of electrons is a chlorophyll molecule that is activated when it absorbs light energy. In the case of mitochondria and bacteria, the source of electrons is the citric acid cycle, and the acceptor of electrons is oxygen. Furthermore, the membranes of mitochondria, chloroplasts, and bacteria somehow use the energy made available by electron transport to synthesize ATP from ADP and phosphate. For this reason, they are referred to as coupling membranes because ATP synthesis is coupled to electron transport. Finally, it was generally agreed that ATP must be synthesized by a chemical process in which one or more of the electron-transport proteins was phosphorylated and then activated to a higher energy state when it accepted an electron. The phosphate could then be transferred to ADP to synthesize ATP. Something like this does in fact happen during glycolysis, so it seemed like a very plausible mechanism at the time. We will refer to this as the chemical hypothesis.

What Mitchell proposed was truly radical. He suggested that the actual function of electron transport was not to activate a phosphorylated protein, but instead to pump protons across the membranes. As a result of the pumping of positively charged protons, a proton gradient was produced across the membranes, which could take the form either of an actual voltage of about 0.2 volt or a pH gradient of about 3 pH units. Because concentration gradients of ions and other solutes provide the energy of osmosis, Mitchell coined the term "chemiosmosis" to describe his concept.

Mitchell also proposed that there was a second enzyme in the membrane that could use the energy of ATP hydrolysis to pump protons. From the laws of thermodynamics, which state that every chemical reaction is in principle reversible, Mitchell stated that this enzyme, now called an ATP synthase, could be driven in the reverse direction if it were embedded in a membrane that maintained a proton gradient going in the opposite direction. As a result, ATP would be synthesized.

Mitchell published the basic concept of chemiosmosis in a short paper in *Nature* in 1961. No one paid the slightest attention. But like all good hypotheses, the chemiosmotic hypothesis had a set of predictions that could be tested experimentally, which would falsify the hypothesis if any one of them failed. There was also the alternative hypothesis: that ATP was synthesized by chemical reactions having nothing to do with proton pumps and reversible ATPases.

The scientific process works best when there are two or more alternative hypotheses, each with predictions that can be tested by critical experiments. The main prediction of the chemical hypothesis was that there should be a phosphorylated protein produced during ATP synthesis—most likely one of the electron transport enzymes. However, no one had been able to demonstrate such an intermediate, which was one of the reasons Mitchell proposed a novel mechanism.

Here are five basic predictions of the chemiosmotic hypothesis:

- Coupling membranes must be able to a maintain proton gradient, and the gradient must be large enough to drive ATP synthesis.
- Coupling membranes must be able to pump protons by electron transport using either light or respiration as an energy source.
- Coupling membranes must contain an ATPase that is also a proton pump.
- Anything that produces a proton leak in the membrane should be an uncoupler. In other words, electron transport will still continue, but ATP will not be synthesized.
- It should be possible to isolate a proton pump and the ATPase and reconstitute them in a lipid vesicle. ATP should be synthesized when the pump is turned on.

I was completing my PhD in the early 1960s while all this was happening, and then I became a post-doctoral researcher at the University of California, Berkeley, where I first became aware of Peter Mitchell and chemiosmosis. Thomas Kuhn was teaching at the Berkeley campus between 1961 and 1964, during which time he wrote and published a book called *The Structure of Scientific Revolutions* in which he brought out his seminal idea of paradigm shifts in science. I doubt that he realized that a paradigm shift was about to happen, and over the next 10 years, I was fortunate to witness this process firsthand.

Over the decade, sometimes by chance, Mitchell's predictions began to be fulfilled. The first was the 1963 discovery by André Jagendorf at Brookhaven National Laboratory that when a suspension of chloroplasts isolated from spinach leaves was illuminated, the pH of the suspension shifted toward a higher pH, and when the light was turned off, the pH returned to its original value. This could be explained if the chloroplast membranes were pumping protons inward, as chemiosmosis required. Two years later, Jagendorf reported that he could synthesize ATP in the dark simply by bathing the chloroplasts in a buffer solution at low pH, then quickly adding a high pH solution containing ADP and phosphate.

The prediction was that the resulting pH gradient alone could synthesize some ATP, and again the prediction was shown to be correct.

By this time, Efraim Racker at Cornell University had discovered that both mitochondria and chloroplasts had an ATPase that he called a coupling factor because the membranes could only synthesize ATP if the factor was present. Mitchell himself, working with Jennifer Moyle in his laboratory near the village of Bodmin in southwestern England, demonstrated that mitochondria could pump protons outward, in the opposite direction from chloroplasts. He also showed that mitochondrial membranes were a sufficient barrier to proton flux so that a proton gradient could be maintained.

The evidence favoring chemiosmosis continued to accumulate even as leading proponents of the chemical hypothesis published their own papers opposing chemiosmosis. Their argument was that even if proton gradients made some ATP, what was actually happening was that pH shifts simply charged up an unknown protein component with phosphate, which was then transferred to ADP to make ATP. But Mitchell's critics had a serious problem. If there was such an intermediate, it should be possible to isolate it. Despite numerous attempts, no one had been able to demonstrate that a phosphorylated protein intermediate was associated with ATP synthesis.

What finally turned the tide was a paper published in 1974 in which Racker and Walther Stoekenius at the University of California, San Francisco, tested the prediction that the only requirement for ATP synthesis was a proton pump and an ATP synthase embedded in the lipid bilayer of a self-assembled membranous compartment. They prepared phospholipid vesicles containing only a light-dependent proton pump called bacteriorhodopsin, which Stoeckenius had discovered earlier, and Racker's coupling factor, the ATPase, embedded in the membrane. When the vesicles were exposed to light, protons were pumped by the bacteriorhodopsin and ATP was synthesized. This paper was published in the *Journal of Biological Chemistry*, and the accumulated weight of evidence finally put chemiosmosis over the top. In 1978, Peter Mitchell was awarded the Nobel Prize, and chemiosmosis is now textbook material.

There was one more Nobel Prize to be awarded in bioenergetics, and a share of the award went to Paul Boyer at the University of California, Los Angeles. Boyer proposed that the ATP synthase had two conformations, the first being a high-energy state that avidly took up ADP and phosphate in such a way that ATP was synthesized and then bound to the active site. In order to release the ATP, Boyer proposed that a subunit in the enzyme was rotating at 100 revolutions per second, driven by proton flux through the stem, and the rotation served both to kick out the ATP and to reset the enzyme in its high-energy state. This idea was a complete surprise, and it would have been scoffed at in the 1960s if Mitchell had suggested it, but so much evidence had accumulated by 1980 that it had to be taken seriously. For instance, John Walker was also awarded a share of the Nobel Prize with Boyer for his research at the Medical Research Council's Laboratory of Molecular Biology in Cambridge, England. Walker crystallized the catalytic subunit of the synthase and used X-ray diffraction to establish its structure, which was consistent with Boyer's proposal. In related work, it was also known that motile bacteria such as *E. coli* used rotating flagellae to produce motion, and Howard Berg at Harvard had shown that proton gradients provided the energy that caused flagellae to spin. This made it easier to think that perhaps the ATP synthase was a kind of molecular motor. The most convincing evidence is that a fluorescent label can be attached to the ATP synthase, and rotation of the enzyme can then be visualized using a microscope.

Could chemiosmosis have provided an energy source for the first forms of life? To me, it seems unlikely that a complete chemiosmotic system of electron transport coupled to proton transport and a rotating ATP synthase was the original source of energy. There may have been a more primitive version, as suggested by Arthur Koch, a microbiologist at Indiana University. Certainly at some point, life became cellular, with membranous compartments containing systems of interacting molecules. And as soon as there were membranes, it would be possible to have ion gradients across the membranes that stored energy. Although the evolutionary steps toward ATP synthesis are not yet established, we can use what we know to speculate on how this might have happened. It is clear that at some point, ATP became the energy currency of the cell, perhaps being synthesized by a chemical reaction as it is in glycolysis. At about the same time, cells found a way to use chemical energy to generate proton gradients, and the energy of the gradient could be used to drive transport of other ions, nutrients, and phosphate across membranes, just as it is today. In a separate evolutionary development, an enzyme appeared that could use ATP hydrolysis to pump protons and produce proton gradients. The stage was then set for chemiosmotic ATP synthesis, in which energy is used to produce a proton gradient, and the stored energy of the gradient drives the reverse reaction of the ATP-dependent proton-pumping enzyme to synthesize ATP.

CONNECTIONS

Why is an understanding of energy relevant to the origins of life? For life to begin, there must have been an energy-driven process in which complex polymers were synthesized to "prime the pump" and then associate into molecular systems. In other words, we must look for a way to add free energy to a system of molecules in order to drive them energetically uphill into more complex molecules and to produce order from disorder so that the entropy of the system decreases. For this reason, we need to understand what kinds of energy were available and demonstrate that the energy would be sufficient to drive polymerization reactions.

There were multiple sources of energy available on early Earth. The most abundant was light energy, and virtually all life today depends on the fact that plants found a way to harvest light and turn it into chemical energy. We call plant life autotrophic, meaning that plant cells use light energy in order to synthesize the molecules they need for growth. However, this requires a complex system of pigments and intermediate reactions, so it seems unlikely that light energy was the energy source of the first forms of life. Instead, the first life would be classified as heterotrophic, meaning that it used nutrients already available in the environment rather than synthesizing the compounds required for growth.

In chemistry, change occurs when reactants turn into products. If a reaction is spontaneous, it continues to equilibrium from a higher energy content to a lower energy content so that energy is released, and at equilibrium, no further change can occur. However, energy can also be put into a mixture of reactants in such a way that the reaction is driven "uphill" to produce products that contain more energy than the original reactants. All life today depends on this fact, and for this reason, we need to understand how energy could drive the chemical reactions related to the origin of life.

A second fundamental concept concerns how fast chemical reactions occur. The rate of a reaction is controlled by activation energy, which is a kind of energy hill that reactants must get over before continuing on to form products. Reactions having low activation energy are fast, while high activation energy reactions can be much slower. An important point to understand is that complex molecules can exist in a kinetic trap, meaning that if they can be synthesized quickly, they will be stable for long periods of time because the degradative reactions are slow. Again, life depends on this fact.

The initial energy-dependent reactions required for life to begin were those that produced potential monomers, such as amino acids, from simple mixtures of gases. After monomers were available, the next requirement for energy was to remove water molecules from between monomers to form chemical linkages, such as peptide bonds and ester bonds, to produce polymers. Life today uses a system of linked reactions called metabolism, but on early Earth, metabolic systems did not exist. Therefore, what was required was another kind of reaction that could produce random polymers and then capture the polymers in compartments called protocells. Each protocell was a kind of natural experiment, testing different sets of polymers that had the potential to undergo catalyzed growth and reproduction. Life began when a system of encapsulated polymers was able to capture energy and nutrients from the environment and then use that energy to grow and reproduce.

7

SELF-ASSEMBLY AND EMERGENCE

It is the closure of an amphiphilic bilayer membrane into a vesicle that represents a
discrete transition from nonlife to life.

HAROLD MOROWITZ, 1992

In the early 1960s, it was becoming clear that all cells had a very thin boundary membrane that formed a permeability barrier between the inside and outside, but no one was quite sure about its molecular structure. There was evidence that cell membranes were about 5 nanometers thick, in the range expected for a lipid bilayer. About this time, J. David Robertson at Duke University began to study cells with a new kind of microscope that used electrons rather than light. The electron microscope had much higher resolving power and could produce images of structures having molecular dimensions. What Robertson saw was astonishing. Cells were not just bags of protein and nucleic acids. Instead, virtually all major components of cells, including the outer plasma membrane, mitochondria, the nuclear membrane, lysosomes, vacuoles, and endoplasmic reticulum, were defined by membranous boundaries. Robertson proposed a model called the "unit membrane" in which he suggested that the membranes were lipid bilayers with functional proteins attached to the surface. But how cells made membranes remained a mystery, and the idea that lipid bilayers were permeability barriers was still just a guess. It was hard to believe that a membrane just two molecules thick could act as a substantial barrier, but in 1963 Alec Bangham proved that it did just that.

Bangham spent his entire research career at the Institute of Animal Physiology (now a much grander Babraham Institute) just outside Babraham village, a few miles south of Cambridge, England. His lab was housed in a barracks-like building, hastily erected during the war and then given over to studies of veterinary disease. I was finishing graduate school in 1965 when I began to hear about "multilamellar smectic phospholipid

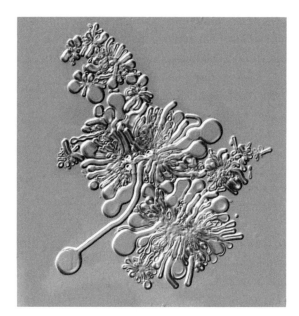

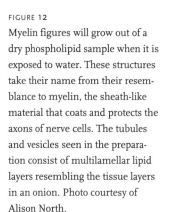

mesophases," the highly technical, tongue-twisting phrase that Bangham originally applied to his discovery. Gerald Weissman, among the first visitors to the Bangham lab, coined the term "liposome" and that came into general usage. It had been known for years that a phospholipid called lecithin could be extracted from egg yolks. (You can still buy lecithin at health food stores as a nutritional supplement, but this is extracted from soybeans.) If a little lecithin was dried on a microscope slide and then exposed to water, long wormlike structures called myelin figures could be seen growing out of the dried lipid (Figure 12). Alec and his coworkers found that the figures had a special quality. When a dilute salt solution was added to egg lecithin in a test tube and then shaken by hand, a milky suspension was produced that consisted of immense numbers of cell-sized spherical globules.

In 1961, Bangham's Institute had just purchased an electron microscope, which uses an electron beam rather than light to illuminate biological specimens, as described earlier. The electron microscope was extremely important in elucidating the internal structure of cells, and it is worth saying a bit more about how it works. Instead of glass lenses, an electron microscope has three circular magnetic lenses that focus the electrons. The lenses (basically electromagnets with a hole in the center to let the electrons through) are located in a vertical tube about 6 inches across and 3 feet tall that rises from a desk with multiple dials and knobs. These control the flow of electrical current through the lenses so that the electrons can be aligned into a beam and focused on the specimen. The source of electrons is a glowing tungsten filament at the top of the tube, to which a voltage of 50,000 volts is applied. The voltage pulls electrons off the filament and accelerates them down the tube, through the specimen, and onto a fluorescent screen or photographic plate where they produce an image.

Colored dyes are used in light microscopy to visualize structures in otherwise clear specimens, but heavy atoms like uranium, lead, osmium, tungsten, and molybdenum are used as stains to provide contrast in electron microscopy. When electrons pass through the specimen, they are scattered by the metal atoms so that a kind of shadow image appears on the screen. To take a picture, the screen is lifted out of the way to allow the electrons to fall onto a photographic plate, which is later developed.

Alec wanted to try out the new instrument, but needed something to look at, so he decided to use the microscopic globules produced by lecithin in water as a first test. After the suspension was prepared, uranyl acetate was added to the mix as a stain, and a small amount was dried on a thin film of plastic supported by a copper grid. The grid was placed in the specimen holder, which was then inserted into the microscope. All the air was pumped out of the column to make a vacuum (electrons don't travel very far in air), the beam was turned on and focused, and the screen began to glow with an eerie green phosphorescence wherever electrons passed through the specimen. Within the green glow, dark globs of phospholipid could be seen outlined by the uranium atoms, as expected. The microscope worked. But then Alec looked more closely at the globs because a strange pattern was appearing. Within the globs, very fine layers were barely visible, looking something like the concentric layers of an onion (Figure 13). Each layer was 5 nanometers thick, just about the same thickness that Robertson and others had observed in cell membranes. This was a revelation: The globs were composed of multilayered membranes, and self-assembled bilayers of a pure lipid had been visualized by electron microscopy for the first time.

Alec and his coworkers went on to demonstrate that compartments within the lipid bilayers could capture solutions of potassium chloride inside the vesicles. In other words, just as they do in real membranes, the lipid bilayers were a barrier to the free diffusion of ions, which could remain in the tiny structures for days. Over the next 30 years, this property made liposomes into a minor industry. They are now used in pharmaceutical

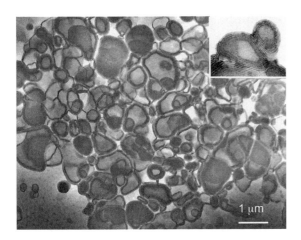

1 μm

FIGURE 13
Lipid vesicles (liposomes) self-assemble when a phospholipid is dispersed in aqueous solutions. The inset shows a multilamellar structure similar to that first documented by Alec Bangham. The layers are produced by onion-like stacks of lipid bilayers 5 nanometers thick.

preparations as a kind of microscopic pill to deliver drugs for treating cancer and fungal infections. They are also essential components of cosmetic preparations like Capture, a cosmetic produced by the Christian Dior company and advertised as something that could temporarily reduce wrinkles in aging skin. In honor of Alec's discovery, the Dior company once flew Alec and his wife Rosalind to Paris on a private jet airplane for a tour of the factory where Capture is manufactured. Alec was also recognized by his colleagues by election as a Fellow of the Royal Society.

THE MOLECULAR FORCES DRIVING SELF-ASSEMBLY

In order to understand how life can begin, we need to know about four basic self-assembly processes and the physical forces that cause them to occur. But first, I want to describe a toy that will give a sense of what happens at the molecular level, where the action is. As children, most of us have played with LEGO® blocks, so we have an intuitive feel for the way that complicated structures can be constructed by popping small round pegs on one block into the holes of another and thus, building recognizable houses, cars, and airplanes from what are basically rectangular bricks. In the analogy to biochemical molecules, the blocks represent monomers such as amino acids, nucleotides, and simple carbohydrates, and the houses, cars, and airplanes are the polymeric proteins, nucleic acids, and starch composed of the monomers. The pegs and holes represent the chemical bonds called covalent bonds that were described in Chapter 4.

The most obvious way to increase the complexity of an encapsulated system of simple molecules is to link them into polymers held together by covalent bonds. By definition, the energy-dependent linking of monomers into polymers is the fundamental growth process of all forms of life, so for life to begin, there must be a process by which polymerization reactions can be driven. Most of the covalent bonds of life are produced by removing water from between monomers to produce ester bonds of lipids, glycoside bonds of starch and cellulose, peptide bonds of proteins, and phosphodiester bonds of nucleic acids. However, the same bonds that are synthesized by removal of water can be broken by addition of water, which is a spontaneous process. Therefore, the first forms of life must have had a way to drive energetically uphill synthesis reactions faster then the downhill reactions of hydrolysis.

Covalent bonds are the strongest forces holding atoms together in molecules and also link molecules into polymers, such as the proteins composed of amino acid chains. It is important to understand that polymer synthesis is not considered to be a self-assembly process. Polymerization does not happen spontaneously; it requires an input of energy. Again, by analogy to LEGO® blocks, you can shake a box full of blocks as much as you want, but they remain individual blocks. Only when you expend energy by pushing pegs into holes can you produce chains of blocks. This is what a living cell does. It gets blocks called nutrients from the environment, rearranges the pegs and holes by a process called metabolism, and then expends energy to link them together into the polymers of life. The overarching goal of research on the origins of life, and the focus of this book, is to understand how this process could have started up in the sterile environment of prebiotic Earth.

Now, let's get back to chemistry. As described in Chapter 4, examples of single bonds are the links holding carbon atoms together in hydrocarbon chains of fats. If only single bonds are present, we call these saturated fats, but if one or more double bonds are present in the center of the chain, the fats are unsaturated, and this changes the properties of the fats. Saturated fats have higher melting points so that the fat we call lard is solid at room temperature. The double bonds of unsaturated fats have a kind of kink in the middle of the hydrocarbon chain, so a fat like vegetable oil is fluid at room temperature. Naturally, I would not go to all the trouble of explaining this unless it were important for understanding life, and later we will see that cell membranes must contain hydrocarbon chains with properties that make them a two-dimensional fluid. Life could not begin or continue if cell membranes were solid.

Let's check our LEGO® blocks again. Imagine that every block not only has the holes and pegs representing covalent chemical bonds, but each also has little magnets embedded in the plastic, one per block. If we put the LEGO® blocks into a box and shake the box vigorously, most of the blocks will spontaneously find their opposite poles in other blocks and stick together in pairs and even chains. The magnetic forces are much weaker than the pegs and holes, so they are continuously being formed and broken, but the blocks are no longer present just as individual blocks. Instead, a variety of more complex structures is produced.

Just as the pegs and holes represent strong covalent bonds, the magnetic interactions are analogous to weaker forces called hydrogen bonds, van der Waals interactions, and electrostatic interactions. These are weak physical forces compared to the strong electron-sharing forces that stabilize covalent bonds, but they are responsible for some of the most important molecular properties of life, particularly self-assembly processes. For instance, hydrogen bonds stabilize the fluid property of water, the double helix of DNA, and the alpha helix of proteins. If there were no hydrogen bonds, water would be a gas, DNA could never replicate, and proteins would remain a tangled mass of random strings. Life as we know it would be impossible.

There is one last important interaction, which is not exactly a force, but is essential to understand how life can begin. Imagine our LEGO® blocks with magnets, but now let's mix in some blocks that lack magnets and shake the box again. It's easy to see that the blocks with magnets will combine into complex structures, but those without magnets will be excluded and forced together into aggregates. This, by way of analogy, is why oil and water don't mix. Water molecules are able to form hydrogen bonds with each other, while oil molecules can stick together only with weaker forces called van der Waals interactions, so they separate into water and oil even if perfectly mixed. But now suppose that we take long LEGO® blocks and include a magnet only at one end. The magnet sticks to the magnets of water, but the non-magnetic end is excluded from water. The molecular versions of such blocks are called amphiphiles. Just as an amphibian like a frog spends part of its life in water and part on land, the water-loving (hydrophilic) head groups of amphiphilic molecules are attracted to water, while the water-hating (hydrophobic) tails are forced together. This physical fact of life, dubbed the hydrophobic effect by Duke University professor Charles Tanford in his 1975 book with that title, is as essential to the molecular system we call life as catalytic proteins and replicating nucleic acids. The hydrophobic effect underlies the ability of amphiphilic molecules called lipids to self-assemble into membranous structures. For cellular life to begin, there must have been a source of such molecules on early Earth.

THE FORCES OF SELF-ASSEMBLY

The four forces that hold life together are covalent bonds, hydrogen bonds, electrostatic interactions, and the hydrophobic effect. Now we will describe these forces individually and try to understand how they could be assembled into a system of molecules that has the properties of life. This, of course, is the central problem addressed by this book: How can something as complex as a living cell separate itself from its surroundings and begin to grow and reproduce? Not only must we think about the separation process, we must place it in the context of the prebiotic environment, in which a variety of energy sources was available to interact with mixtures of organic solutes and ions in aqueous solutions. Out of this mixture, the earliest forms of life needed to extract specific small molecules, bring them across the cell boundary, and then use energy to incorporate them into growing polymers. Finally, there must have been a mechanism by which the cell, having grown to a certain size, was able to divide into smaller versions of itself, which could then begin the growth cycle again.

Now we must learn a little more about how self-assembly actually works. Unless you have taken a college chemistry course, words like hydrogen bonding, entropic effects, electrostatic interactions, and van der Waals forces will sound like a foreign language. But think how impressive you will be at the next dinner party when you explain that mayonnaise is actually an oil-in-water emulsion stabilized by entropic effects of amphiphilic phospholipids present in the egg yolks used to make it. My friend Harold Morowitz even wrote a book of essays with the title *Mayonnaise and the Origin of Life*.

HYDROGEN BONDS STABILIZE BIOMOLECULAR STRUCTURES

When thinking about how forces work at the nanoscopic scale, it's always good to begin with a few examples with which everyone is familiar. Let's start with rotten eggs, water, and ice. The distinct aroma of rotten eggs is nothing more than a sulfurous relative of water called hydrogen sulfide. (Water is hydrogen oxide.) Hydrogen sulfide (H_2S) is produced by bacteria that use sulfur-containing amino acids as a source of nutrients and energy, but they can't use all of the sulfur, so it gets dumped in gaseous form. The question is: Why is H_2S a gas while H_2O is a liquid, even though sulfur is a heavier atom than oxygen? Keep that question in mind. Now let's think about the difference between water and ice. At 1 degree above its freezing point, water is a liquid, but at 1 degree below, it is a solid. What happens when water becomes ice? The answer to both of the above questions has to do with the first force of self-assembly, which is called hydrogen bonding. This is the force acting between water molecules that holds them together in a liquid. Hydrogen sulfide cannot form hydrogen bonds, so it is a gas. Just above the freezing point, water is held together in liquid form by hydrogen bonds—but as small clusters that are constantly forming and breaking up. What happens when water freezes is that the hydrogen bonds become continuous throughout the water, and the water becomes a solid.

Linus Pauling was first to recognize how hydrogen bonds can drive self-assembly processes in proteins, one of the most important biological polymers. In 1939, as a professor of chemistry at Cal Tech, Pauling published his classic book *The Nature of the Chemical Bond*. In it, he presented the first clear and comprehensive explanation of the kinds of forces that stabilize molecular structures, including hydrogen bonds. World War II interrupted Pauling's basic research, but in 1948 he had a brilliant insight into how hydrogen bonds could stabilize a protein chain into a beautiful helical structure he named the alpha helix and into planar structures now called beta sheets. These structures arise spontaneously as protein chains are synthesized by ribosomes and are dominant features of most proteins. For his discoveries, Pauling received the Nobel Prize in 1958.

The other fundamental biological structure stabilized by hydrogen bonds is the double helix of DNA. In the early 1950s, Pauling began to think about solving this structure as well and published his best guess, a triple helix. At the same time, two young graduate students working at Cambridge University in England decided to solve the same problem. The students were James Watson and Francis Crick, who by a mixture of critical thinking, model building, and good luck were able to present a correct double helix structure in 1953 in the journal *Nature*. In Watson and Crick's structure, adenine paired with thymine with two hydrogen bonds, and guanine paired with cytosine with three hydrogen bonds. These are now referred to as Watson-Crick base pairs or more generally, as complementary base pairings. Toward the end of their one-page paper, Watson and Crick indulged in a classic bit of British understatement: "It has not escaped our notice that the specific pairing we have postulated immediately suggests a possible copying mechanism for the genetic material."

And that's exactly right. Somehow, over three billion years ago, the first forms of life discovered the same secret, that hydrogen bonding complementarity allowed one strand of DNA to be copied into a second strand. How this could have happened is perhaps the central mystery of the origins of life. Later in this book, I will propose a possible solution.

A final question about the hydrogen bonds that stabilize the alpha helix and beta-sheet structures of a protein and the double helix of nucleic acids: Why are these classified as forces of self-assembly, while this term is not usually applied to covalent bonds? The reason is that they occur spontaneously and can add order to an otherwise disordered system of molecules, even though they are much weaker than covalent bonds. Imagine that we have a protein molecule and a nucleic acid molecule dissolved in water and that we heat the water to near boiling temperature. Under these conditions, the alpha helices, beta sheets, and double helices fall apart because the hydrogen bonds are not strong enough to hold them together against the violent molecular motions produced by the heat. The protein, originally globular, becomes a disordered string, and the DNA double helix falls apart into single strands. Now let's cool down the hot solutions. Something pretty amazing occurs. One by one, the hydrogen bonds in the protein chain begin to find each other and spontaneously produce all the original alpha helices and beta sheets. The complementary base pairs of the DNA strands also begin to line up as adenine forms hydrogen bonds with thymine, and guanine with cytosine. After several minutes, or even several hours, the complete double helix has reassembled.

The take home message is that hydrogen bonding is essential for the functions of cellular life today. In order to discover how life began, we need to understand how hydrogen bonds could have added spontaneous order to organic compounds and their polymers in the prebiotic environment.

VAN DER WAALS FORCES

Let's think about van der Waals interactions next. They were discovered by Johannes Diderik van der Waals, who was born in Leyden, The Netherlands, in 1837. Johannes was doubtless a genius-level intellect, but with limited opportunities, he first became a schoolteacher with just an elementary school education. However, he had a passion for science and persisted in educating himself, finally enrolling in Leyden University and completing a doctorate in 1873. His genius became apparent with the publication of his first paper, in which he derived equations that described the physical properties of gases and liquids. This and several later papers had an enormous impact, leading other scientists to realize that even gases like hydrogen and helium could be liquefied. And a few years later, they were. For his discoveries, van der Waals was awarded the Nobel Prize in 1910.

Van der Waals' equations predicted that gases are not hard little spheres like billiard balls, but instead have a kind of stickiness that produces a weak force acting between them. This force is analogous to the way a bit of dust will jump onto a comb if the comb is electrified by rubbing it with a woolen cloth. The comb gains an electrical charge by literally rubbing extra electrons off the wool, what we commonly call static electricity, and this induces an opposite charge on the surface of the dust when the charged comb approaches. Opposite charges attract, and the dust leaps onto the comb. Something like this happens at the atomic and molecular levels as well. The electronic shells around atoms are not perfect, but instead undergo extremely fast fluctuations. The transient electrical charges induce the opposite charge in neighboring atoms and molecules. The resulting attractive forces are now called van der Waals interactions.

What does this have to do with biology, or for that matter, the origins of life? The main point is that van der Waals interactions are the only forces that act between chains of hydrocarbons, so they determine whether hydrocarbons will be gases, liquids, or solids at any given temperature. For instance, methane, ethane, and propane contain one, two, and three carbons respectively, and they are all gases at ordinary temperatures and pressures. But if you look closely at a plastic cigarette lighter, you will see a clear liquid in the container. This is butane, with four carbons, and it is just on the edge of being a gas or a fluid. The lighter is cleverly designed so that when you use your thumb to turn the small wheel that makes a spark, a valve is opened and the butane turns from liquid to gas and ignites.

The next five hydrocarbons from pentane to decane are all fluids at ordinary temperatures, which is why gasoline is a liquid. But when the chain lengths increase further, to the undecane and dodecane that sit on a shelf in my laboratory (11 and 12 carbons) they are solids in their bottles unless it is a warm day.

What is happening? Why this transition from gas to liquid to solid? The reason is that as chain length increases, the van der Waals forces acting on any single molecule are additive. Imagine a zipper with just one, two, or three teeth. It would not hold anything together very well. But the zipper works increasingly well as the number of teeth increases from five to 10, and with 11 or more teeth it turns into a fairly strong fastener. Later on, I will use this analogy again to describe how the DNA double helix comes apart into two separate chains, a process that we call unzipping, for lack of a better word.

ELECTROSTATIC FORCES

Most of us have seen what happens when bacteria called lactobacilli begin to multiply in milk. The milk "goes sour" and produces white clumps. Often this occurs by accident when milk is left out of the refrigerator and gets warm enough for the bacteria to grow, but certain strains of lactobacilli are also used to produce food items like sour cream, yogurt, and a variety of cheeses. But why does the milk change from a white liquid to large clumps? The answer has to do with electrostatic forces. There is a protein in milk called casein, which at neutral pH ranges is negatively charged. Proteins are composed of 20 different amino acids, most of which are uncharged, such as glycine, alanine, serine, and valine. But two of the amino acids—aspartic acid and glutamic acid—have an extra carboxylate group ($-COO^-$) that is negatively charged, and two others—lysine and arginine—are positively charged by an extra amine group ($-NH_3^+$). Most proteins have more anionic than cationic amino acids, which gives them a net negative charge. This is essential because the negative charges keep the proteins apart in solution. What happens in milk is that the lactobacilli produce lactic acid, the "sourness" that we can taste. The acid is due to hydrogen ions, and the pH of the milk goes from neutral to acidic pH range, around pH 5. The hydrogen ions bind to the carboxylate groups by the following reaction:

$$H^+ + R–COO^- \{\Rightarrow\} R–COOH$$

(The letter R is just an abbreviation used by chemists to indicate an unspecified chemical group.) The main point is that the negative charge disappears. The casein molecules are no longer repelled, so they begin to adhere to each other, resulting in large clumps of protein.

For the most part, cells work hard to maintain their internal pH in the neutral range in order to keep proteins negatively charged and prevent clumping. There is an important exception, and these are the proteins called histones that have a lot of lysine and arginine in their structure, and therefore, are positively charged. The positively charged histones interact strongly with negatively charged DNA in the nucleus of cells, and the resulting structure, called chromatin, stores genetic information in the nucleus until it is needed to direct protein synthesis.

What role could electrostatic forces play in the origins of life? As we will see later in this chapter, interactions between ions and organic molecules can promote or inhibit self-assembly processes. I will just give a few examples here. Take phosphate, for instance, which is an essential component of all life today. Phosphate is anionic, which means that it is negatively charged at ordinary pH ranges. Calcium, however, is a cation with two positive charges (Ca^{2+}) in solutions. By themselves, both phosphate and calcium ions are soluble in water. Sea water, for example, has 10 millimolar calcium in solution along with sodium chloride that makes it salty. But when calcium and phosphate are both present as ions in water, their positive and negative charges interact in such a way that a virtually insoluble crystalline mineral called apatite precipitates. This is fine if you are building a tooth or a bone (both are hard because they contain apatitite crystals), but it's not so good if you are trying to use phosphate in one of the reactions related to the origins of life. How phosphate became an essential component of metabolism remains an open question.

Another example is the assembly of membranes from amphiphilic molecules like fatty acids. When you wash your hands with soap, the slippery feel is actually produced by membranes of soap molecules that easily slide past one another and act as a lubricant. But just try washing in sea water with its high calcium and magnesium content. The soap coagulates into hard curds and loses its ability to dissolve grease. It also loses its ability to self-assemble into membranous vesicles, as we demonstrated in 2002 in a paper I co-authored with Pierre-Alain Monnard, now at the University of Southern Denmark, and Charles Apel at San Jose State University. To me, this simple fact could be a powerful constraint imposed on the site where life began because it would seem to exclude the ocean. Furthermore, most of my colleagues would not choose sea water to run their simulation experiments, but instead would use very pure laboratory water. Perhaps we should also think about the origin of life in a freshwater setting, rather than a marine environment. By the way, well water that contains calcium and magnesium is referred to as "hard water." This water can be filtered through a water softener, which exchanges sodium ions for calcium and magnesium ions. The sodium, having only one positive charge instead of the two charges of calcium and magnesium ions, does not form insoluble curds with negatively charged soap molecules, so the hard water is softened.

The last self-assembly process affected by electrostatic forces concerns mineral surfaces. Common minerals composed of silicates, such as quartz, have only weakly charged surfaces, but another common mineral contains aluminum and is called clay. Microscopic particles of clay have charged surfaces that strongly bind organic compounds with ionized groups. As we will see later, clay minerals have played an important role in research on the origins of life because the organic compounds that bind to the clay surfaces can undergo polymerization reactions.

HYDROPHOBIC EFFECTS DRIVE SEPARATION OF MOLECULAR SYSTEMS FROM THE ENVIRONMENT

Life cannot exist as a solution of molecules, no matter how complex, because molecules in solution diffuse randomly and are unable to maintain the organized state required for cellular functions such as metabolism, catalyzed growth, and reproduction. For this reason, all life today is cellular, and cells are bounded by amphiphilic molecules that self-assemble into membranous structures. The membranes are stabilized by a unique interaction between hydrocarbon chains of the amphiphiles and water, which is called the hydrophobic effect. In order to understand how this works, we will first need to have a better understanding of the kinds of molecules that compose membranes.

As described in Chapter 4, the most common amphiphilic molecules incorporated into the membranes of living cells are called fatty acids. Their common names give a hint of their biological source, and their technical names say something about the number of carbon atoms in their structure. For instance, a fatty acid with 16 carbons looks like this:

$H_3C–CH_2–CH_2–CH_2–CH_2–CH_2–CH_2–CH_2–CH_2–CH_2–CH_2–CH_2–CH_2–CH_2–$
$CH_3–COOH$

This fatty acid was first isolated from palm oil and is called palmitic acid, while its technical name is hexadecanoic acid. Myristic acid (tetradecanoic acid), with a 14-carbon chain, is from the oil of the nutmeg tree (Latin name *Myristica*); lauric acid (dodecanoic acid), with 12 carbons, is from oil of the laurel bush; and capric acid (decanoic acid), with 10 carbons, was so named because its strong aroma is reminiscent of goats. Sometimes the common names of hydrocarbon derivatives make it into the commercial realm. For instance, in 1898, Mr. B.J. Johnson made a soap from a mixture of palm oil and olive oil, which contained palmitic and oleic acid. You can still buy the soap. It's called Palmolive. (Oleic acid has 18 carbons and a single double bond in the middle of the chain.)

As described earlier in this chapter, the self-assembly of membranous vesicles is a spontaneous process in which amphiphilic compounds form vesicles bounded by bilayer structures. Because the hydrocarbon chains of the amphiphiles form an oily layer in the center of the bilayer, ions and polar solutes cannot readily diffuse across the bilayer. This means that the vesicle's internal volume is separated from the external environment by a barrier which can capture mixtures of large and small molecules inside. Each vesicle is different from all the rest, and each vesicle therefore represents a kind of natural experiment that addresses this question: What happens when this particular mixture of molecules is brought together in a confined space? Recent attempts to answer that question are driving much of the current research on the origins of life, and will be discussed in later chapters.

WHAT WERE POSSIBLE SOURCES OF AMPHIPHILES ON PREBIOTIC EARTH?

It should be clear by now that I am a great fan of amphiphilic molecules. Amphiphiles form the boundary membranes of all cellular life today, and later in this book, I will argue that membranous structures were also essential for life to begin. At this point, it will be helpful to summarize what we know about possible sources of membrane-forming amphiphiles on early Earth. In the first chapter of this book, I related how we discovered amphiphilic compounds in the Murchison meteorite and demonstrated that they could form membranous vesicles. It follows that a possible source of such compounds is that they were delivered to the young planet during late accretion following the moon-forming event. I also described my work with Jason Dworkin, Scott Sandford, and Lou Allamandola at the NASA Ames Research Center, where we found that amphiphiles were produced when UV light interacted with thin layers of ice containing methanol, carbon monoxide, and ammonia. This reaction, which occurs on dust particles around stars, probably accounts for at least some of the organic compounds of carbonaceous meteorites, including the amphiphilic material. Then there is the Fischer-Tropsch synthesis in which carbon monoxide binds to a hot metal surface and reacts with hydrogen. As a result, the carbon monoxide is reduced to methylene ($-CH_2-$) and forms long hydrocarbon chains. If the chains have a carboxyl group ($-COOH$) or hydroxyl group ($-OH$) at one end, they are called fatty acids or alcohols. We have found that mixtures of the acids and alcohols readily assemble into beautiful membranous vesicles.

Now I want to describe one other possible source of amphiphiles. Bob Hazen works at the Carnegie Institution of Washington, where he carries out basic research on the way that organic compounds interact with mineral surfaces. He is particularly interested in what happens to organics in the high temperatures and pressures associated with volcanic sites. Bob wrote to me several years ago, expressing interest in my research and proposing a collaboration in which we see whether any amphiphiles are synthesized in simulations of volcanic conditions.

The way that Bob performs the simulation is both interesting and dramatic. He uses a gold capsule about 1 inch long, which has been welded shut. Gold is used because it is chemically stable and will not affect the reacting compounds of their products. A compound of interest is sealed inside the capsule, which is then heated to 250°C for a couple of hours. As the compound reacts, it gives off gas, and the pressure in the capsule increases to 2,000 atmospheres. This is such a high pressure that it is close to being explosive, so when Bob opens the capsule to sample the reaction products, he first must cool it in liquid nitrogen to freeze the gas and lower the pressure. Even with this precaution there is an audible pop when he cuts off one end with a pair of wire cutters.

We could have used almost any organic compound for the experiment, but we chose to use pyruvic acid. A biochemist will tell you that pyruvate is central to all metabolism. It is the end product of glycolysis, in which glucose is enzymatically broken down to smaller molecules and then enters the mitochondrion where it releases its energy content during oxidation to carbon dioxide. Bob and his colleagues at the Carnegie Institution had shown that pyruvic acid could be synthesized in simulations of volcanic conditions, so it seemed like a good choice to start our experiment. Bob prepared several of the gold capsules in his lab and then brought them to Santa Cruz for us to test.

The next morning, we were ready to go. One of the gold capsules was dropped into a bath of liquid nitrogen, producing a burst of misty gas as it quickly cooled to −180°C. Bob reached down into the boiling liquid nitrogen with the wire cutters, and there was a popping sound and another cloud of mist as the pressure was released. This was followed by the distinct sweet aroma of caramel candy! He dropped the capsule into a mixture of chloroform and methanol to dissolve whatever amphiphiles might have been formed by the reaction, and the solvent immediately turned a golden brown. Pyruvic acid is a white crystalline compound that would not be soluble in an organic solvent like chloroform, so whatever was produced during the reaction was very different in its physical properties.

Now it was my turn. I dried some of the brown solution on a microscope slide, added some water and a cover slip, and examined the slide with my microscope. Something pretty amazing occurred as the water began to interact with the product. Pyruvic acid itself is highly soluble in water, but this stuff did not dissolve. Instead, it expanded, forming a kind of foamy structure that finally filled the slide. The material was highly fluorescent, so I took a fluorescence micrograph (Figure 14).

The lesson we learned from this work and other experiments is that there are numerous ways for amphiphilic molecules to be synthesized. For me, this means that they

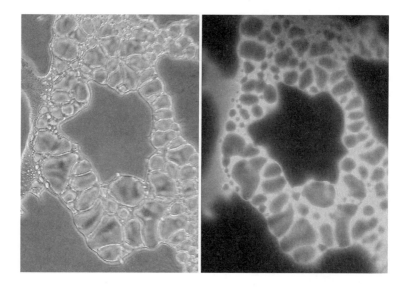

FIGURE 14

Pyruvic acid undergoes a complex series of reactions when it is exposed to elevated temperature and pressure, simulating the geothermal conditions of volcanic activity. Some of the products synthesized are amphiphilic molecules that can assemble into membranous compartments (left panel). Other products are fluorescent, and the panel on the right shows the same material with its natural fluorescence activated by ultraviolet light.

were likely to be abundant constituents of the prebiotic mixture of organic compounds. Their ability to assemble into membranous compartments means that cell-sized vesicles would be common on early Earth, ready to serve as homes for the first forms of cellular life.

COULD SELF-ASSEMBLY PROCESSES ACCOUNT FOR THE ORIGIN OF LIFE?

Given the self-assembly forces described above, the basic theme of this book is that such forces, in combination with certain kinds of energy, led to the emergence of a simple form of cellular life. It's not sufficient, of course, just to make a claim like that. The next step is to see whether it represents a useful hypothesis. If so, predictions will arise that can be tested experimentally in the laboratory. I will discuss this possibility in detail in Chapter 14, but here I will sketch the basic ideas to show how self-assembly can produce structures we refer to as protocells, which are membranous compartments containing large polymeric molecules that are not alive, but are stepping stones on the path to the origin of life.

The first thing to do is to make microscopic compartments that contain the polymers and allow them to interact with one another, particularly the proteins and nucleic

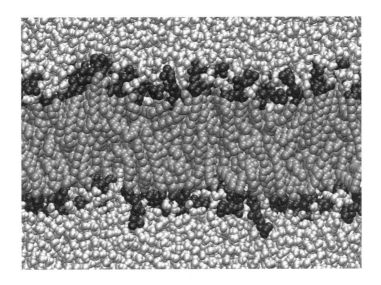

FIGURE 15

Computer simulation of a lipid bilayer. The dark gray material in the center represents hydrocarbon chains of the two lipid layers, surrounded on either side by light gray water molecules. Lipid head groups are shown in black. A few water molecules are beginning to penetrate into the bilayer, and in fact water is able to diffuse across the lipid bilayer membrane. Image courtesy of Andrew Pohohrille.

acids that comprise the machinery of life. As described earlier in this chapter, making vesicles is easy because of the physical properties of lipids, the amphiphilic molecules that self-assemble into bimolecular layers. Although individual lipid molecules are too small to see, even with an electron microscope, we are able to program a computer to give a near approximation. The resulting images are called molecular dynamics simulations, and they can even illustrate the actual motion of the molecules for a few nanoseconds of time. Figure 15 shows a computer model of a lipid bilayer in cross-section that illustrates the two juxtaposed lipid layers. The carbon backbones of the hydrocarbon chains fill the center of the image, surrounded on either side by white water molecules. If you could see the motions of the chains and water in the nanosecond of time simulated by the computer, the chains would be whipping about and spinning, with jostling water molecules jumping in and out of the image. In other words, both the lipids and the water are fluids. Notice that there are almost no water molecules within the hydrocarbon chains. The computer program has simulated the fact that water and oil don't mix. The main point to understand is that whenever phospholipids are present in water, they spontaneously form lipid bilayers.

The next step is to find a way to trap large polymers inside the vesicles. Neutral molecules like water, oxygen, and carbon dioxide that don't have an electrical charge are small enough to dissolve in the lipid bilayer. What happens is that the rapid molecular motions of the hydrocarbon chains are constantly producing holes that can accommodate small molecules, which hop into the holes and diffuse across to the other side. But the electrical charge on common ionic solutes such as sodium (Na^+), potassium (K^+), and chloride (Cl^-) makes it very difficult for them to dissolve in the oily layer, and it is virtually impossible for larger molecules like proteins and nucleic acids to get across. So how could the first cell-like structures have been produced?

There is only one way that seems plausible under the conditions of prebiotic Earth, which is to temporarily break open the membrane barrier to let large molecules in, then reseal it. We discovered years ago that a single cycle of drying and wetting could accomplish this task, a process that will be described in detail in the next chapter on encapsulation. Here we will simply say that if mixtures of phospholipid vesicles and large molecules like proteins or nucleic acids are dried, they form a film. The vesicles flatten out in the film as it dries, then fuse into literally millions of lipid layers that trap the molecules between the layers. When water is added back to the dried film, the lipid reassembles into new vesicles, but now up to half of the large molecules that were originally present outside the vesicles have been trapped inside.

Where could this happen on early Earth? Wetting and drying cycles could occur in many locations, such as intertidal zones along seashores, freshwater ponds, or my favorite, the edges of geothermal springs that would have been common four billion years ago. I am now testing this scenario by traveling to volcanic regions and adding potentially self-assembling molecules to hot springs. As you will see in the next section, this exercise has produced some surprises, but it also suggests that we are basically on the right track.

SELF-ASSEMBLY PROCESSES IN NATURAL ENVIRONMENTS

I have always been bothered by how glibly we experimental scientists assume that what goes on in the laboratory must also take place in the real world. In the lab, we use clean glassware, pure water, and pure chemicals, and we design and perform experiments under carefully controlled conditions of pH, salt concentrations, and temperature. We also try to make experiments as simple as possible in order to reduce variables to a manageable level so that we can understand the results.

And yet, as described in Chapters 2 and 3, early Earth was a hugely complex chemical and physical environment, and organic compounds were anything but pure. This is the reason I traveled to Kamchatka, to Hawaii, to Iceland, and to a bubbling acre of hot springs called Bumpass Hell on the slopes of Mount Lassen in northern California. Now that we have discussed the forces of self-assembly, I can tell the rest of the story and what I learned.

First, it is important to make clear what we are testing. As described earlier in this chapter, at least in the laboratory, we can demonstrate self-assembly of amphiphilic molecules into membranes, and the resulting vesicles can encapsulate large molecules like proteins and nucleic acids. If the macromolecules happen to be enzymes, we can also study the vesicles as model systems with some of the properties of life, such as nucleic acid replication or protein synthesis. We assume that the geothermal environments of modern volcanoes, and particularly the hot springs, resemble the conditions on early Earth at the time of life's beginning. The main difference is that today's atmosphere contains oxygen and nitrogen, but early Earth's atmosphere would have had no oxygen, but instead would be a mixture of carbon dioxide and nitrogen. When we measure the physical and chemical properties of modern hot springs, we find that their temperatures range up to near the boiling point of water. The springs are usually acidic, with pH down around 3, and they contain small amounts of dissolved mineral cations like magnesium, calcium, iron, and even aluminum. They also have deposits of clay minerals, as well as the mineral surfaces of the lava rocks that contain the pools. And last, they are continuously forming and evaporating, sometimes by simple splashing on surrounding rocks by the boiling action, other times by cycles of rain and dry weather.

So, the main question I am asking is whether self-assembly can occur under these conditions. More specifically, how do heat, pH, and ionic composition affect self-assembly processes? Furthermore, if life begins when a system of molecules is able to make polymers like proteins and nucleic acids, can covalent bond formation take place in a geothermal environment? And even if this is possible, how stable are the bonds?

DARWIN'S WARM LITTLE POND

In 1871, Charles Darwin wrote a letter to his friend Joseph Hooker in which he expressed the following idea:

> "It is often said that all the conditions for the first production of a living organism are now present, which could ever have been present. But if (and oh! what a big if!) we could conceive in some warm little pond, with all sorts of ammonia and phosphoric salts, light, heat, electricity &c., present, that a proteine compound was chemically formed ready to undergo still more complex changes, at the present day such matter would be instantly devoured or absorbed, which would not have been the case before living creatures were formed."

Darwin could not have known that he was suggesting a scenario for a prebiotic environment which would still be a viable alternative nearly 140 years later. As discussed in Chapter 2, various alternative sites for the origin of life have since been proposed, ranging from transient melt zones in a global ice pack to hydrothermal vents and hot subterranean mud. Many such sites contain microbial populations that are extremophiles, and it has been suggested that the last universal common ancestor lived in a deep hydrothermal site secure from sterilizing giant impacts. However, a surface site at an intermediate temperature range, exposed to the atmosphere, and able to produce concentrated reactants by evaporation remains a viable alternative for the site of the origin of life. After life began, it could then evolve into a variety of niches, including the extremophile environments.

Darwin might have imagined his pond to resemble the tropical saltwater tide pools he had observed in the Galapagos Islands, or a freshwater pond similar to those that dot the English landscape. He specified a form of fixed nitrogen (ammonium salt) and phosphate, presumably because he knew that nitrogen and phosphorus are among the six major elements of life. He did not include an explicit carbon source or mention any organic compounds other than proteins because in 1871, it was not yet understood that amino acids and nucleobases were the major monomers of life's polymeric macromolecules.

The plausibility of Darwin's pond has never been tested in a natural environment. At a minimum, such a site is expected to have access to liquid water and organic compounds, either synthesized *in situ* or transported to the site. Possible sources of organics include geochemical synthesis or delivery by extraterrestrial sources during late accretion. From the earlier results described above, it is possible that self-assembly of amphiphiles might occur under these conditions. It is even conceivable that synthesis of peptide bonds and ester bonds would be possible. However, the monomers must somehow be exposed to conditions that permit reactions to occur. But if the conditions are such that the monomers are too dilute or precipitate as insoluble salts or tightly adsorb to mineral surfaces, polymerization chemistry would not be possible.

The experiments I will describe were undertaken in geothermal sites in Kamchatka, Russia, to determine how a set of pertinent organic compounds behave in volcanic settings. The Kamchatka Peninsula is just north of Japan, in the far east of Russia, and is nearly the size of California. It is the easiest part of Russia to visit if you are a California scientist. To do so, we board an Alaska Airlines plane in San Francisco, and a few hours later land in Anchorage, Alaska, where we stay overnight in a motel near the airport. Early next morning, we climb into an old jet airplane operated by Magaden Airlines, joining a crowd of hunters and sports fishermen who have discovered the wilds of Kamchatka. This flight runs just once a week, and if you miss it going or coming back you must wait another week.

We land in Petropavlosk, a bustling little city of 200,000 surrounded by volcanic peaks, hinting at the fact that Kamchatka has nearly one-tenth of all the active volcanoes in the world. The city was founded by Vitus Bering in 1740, who named it after his two ships, *St. Peter* (*Petro*) and *St. Paul* (*Pavlov*) in Russian. We are met by a local scientist, Vladimir Kompanichenko, who carefully organized the field work for us and will collaborate in the research effort. After an overnight stay in a local inn, our team of six scientists from Stanford University, the Carnegie Institution of Washington, and the University of California, Santa Cruz, settle into a former Russian Army troop carrier owned by the driver and his wife, who will take care of us for the next week. Five hours later we are grinding up the lower slopes of the volcano, through a landscape that bears a striking resemblance to the images sent back from the rovers on Mars.

Over the next few days, we explored the crater described in Chapter 2, and finally chose a test site on the flank of Mount Mutnovski. This consisted of several acres of boiling hot springs and roaring vents spewing out steam and the distinctive rotten egg aroma of hydrogen sulfide, representing a modern analogue of the kind of geothermal sites that we presume would also be common on prebiotic Earth.

For the experiment I wanted to do, I needed a small, relatively quiet pool containing just a few liters of water. Most of the pools in our site were either too large or had water running through them so that anything that I added would be diluted or washed away. Fortunately, we did find one site that was just about ideal. The pool had a heat source in the center which maintained a near-boiling temperature of 97°C at this altitude. It was lined with a layer of whitish clay mineral several centimeters deep, which was disturbed by boiling to produce a constant agitation and turbidity. In Chapter 4, I described the molecular components of life and the kinds of organic compounds that were likely to be available on early Earth. I used this knowledge to decide what components should be present in the prebiotic soup to be added to Darwin's hot little puddle. These included four amino acids (glycine, L-alanine, L-aspartic acid, L-valine—1 gram each), four nucleobases (adenine, cytosine, guanine, and uracil—1 gram each), sodium phosphate (3 grams), glycerol (2 grams), and myristic acid (1.5 grams). I expected this particular mixture to be a useful guide for monitoring self-assembly of the myristic acid into membranes, as well as serving as a potential source of reactants for any synthetic reactions that might occur. I first took a series of samples that would tell us what was in the water before the experiment began. Then I dissolved the powdered mixture in 1 liter of hot water from the site and poured the milky solution into the boiling water to start the experiment.

The first clue appeared immediately after adding the mixture, when a white precipitate appeared on the surface of the pool and accumulated around the border. Everything else was soluble in water, so this had to be the myristic acid, but at 97°C, it should be melted, not a white precipitate. This was very puzzling. I took water samples at time intervals of minutes, hours, and days, and I also took samples of the clay lining the pool before and 2 hours after the organic mixture was added. We took the samples back to our laboratory in California, where they were analyzed to see what had happened to the organic compounds.

So, what did we find out? The results were totally unexpected. Most of the added solutes rapidly disappeared from solution, some within minutes, others over a period of several hours. The only exception was the fatty acid, which we could still detect nine days after it was added. Where did the stuff go? Fortunately, I had also taken samples of the clay lining the pool. Back in the laboratory, when we made the pH alkaline by adding sodium hydroxide, virtually all of the compounds that we had added reappeared in solution. They had been tightly adsorbed to the clay in the pool, which is why they disappeared from the water samples.

This experiment taught some significant lessons. Perhaps most important was that we should not expect laboratory simulations to predict what would happen when the same experiment is performed in the actual environment that was being simulated. For instance, I expected that the myristic acid would readily disperse in the hot water, but instead it formed an insoluble precipitate. When we analyzed the water from the Kamchatka pond I realized why. The pond water contained small amounts of dissolved aluminum and iron, and these reacted with the myristic acid to produce insoluble curds that lacked membrane-bounded compartments. My unhappy conclusion was that if membranous vesicles were required for the origin of cellular life, they were not going to self-assemble from fatty acids, at least in this particular geothermal environment. And in a laboratory simulation I would never have thought of adding aluminum and iron to the solution, yet these are abundant in a geothermal hot spring and completely inhibit self-assembly processes.

The second hard lesson was that if clay minerals are present, they will adsorb even gram quantities of added amino acids, nucleobases, and phosphate—concentrations much higher than we might expect to be present in the prebiotic environment. Clay has always been considered to be a positive factor in the origin of life, acting to promote synthetic chemical reactions, but Darwin's hot little puddle taught me a different lesson. Clay could be more like a sponge, removing potential reactants from solution before they have a chance to find each other and undergo a chemical reaction. The challenge now, to myself and to my colleagues, is this: If we think a laboratory simulation might be relevant to the origins of life, test the idea in a natural version of the simulation. We might be surprised by the results, which may force us to reconsider our assumptions.

CONNECTIONS

The pathway to the first forms of life must have involved multiple self-assembly processes in which simple mixtures of organic compounds in solution were organized into more complex systems. The underlying forces of self-assembly include hydrogen bonding, electrostatic interactions, van der Waals interactions, and hydrophobic effects. Self-assembly processes are spontaneous, unlike energy-dependent covalent bond formation, but the self-assembly bonds are also weaker and more strongly influenced by environmental conditions such as pH, temperature, and ionic solutes.

In the laboratory, self-assembly can produce membranous compartments from simple amphiphilic compounds (like fatty acids) that were likely to have been available in the prebiotic environment. Such compartments would be essential for the origin of cellular life. However, those same organic compounds in a natural setting would have had a variety of possible fates other than what we observe in the laboratory, where pure compounds react in glass or plastic containers. For instance, we found that self-assembly of boundary structures cannot occur under hot acidic conditions, at least with a simple amphiphile such as myristic acid. Furthermore, if cations are present, such as the iron and aluminum in the Kamchatka pond or the calcium and magnesium in sea water, these cations will precipitate fatty acids as insoluble soaps and inhibit their ability to assemble into membranes.

These considerations are useful constraints on the conditions required for life to begin that were described in Chapter 2. They suggest that, from a biophysical perspective, the most plausible planetary environment for the origin of life would be a moderately acidic aqueous phase in the range of 70°C to 90°C, with divalent cations at submillimolar concentrations. This suggestion is in marked contrast to the view that life most likely began in a high-energy geothermal or marine environment, perhaps even the extreme environment of a hydrothermal vent. The reason a marine site for life's beginning has been favored is simply that fresh water would be rare on early Earth. Even with today's extensive continental crust, fresh water only represents about 1% of Earth's reservoir of liquid water. Another concern about a freshwater origin of life is that the lifetime of freshwater bodies tends to be short, on a geological time scale.

On the other hand, if the temperature and ionic composition of a geothermal spring or sea water markedly inhibit self-assembly processes, we might need to reconsider the assumption that life inevitably began in such environments. A more plausible site for the origin of cellular life may resemble Darwin's inspired suggestion of a warm little pond, such as freshwater pools maintained by rain falling on volcanic land masses, which fluctuate between dry and wet conditions. After the first forms of cellular life were established in a relatively benign environment, life would rapidly begin to adapt through Darwinian selection to more challenging conditions, including the extreme temperatures, salt concentrations, and pH ranges that we now associate with the limits of life on Earth.

HOW TO BUILD A CELL

The early cells were just little bags of some kind of cell membrane, which might have been oily or it might have been a metal oxide. And inside you had a more or less random collection of organic molecules, with the characteristic that small molecules could diffuse in through the membrane, but big molecules could not diffuse out. By converting small molecules into big molecules, you could concentrate the organic contents on the inside, so the cells would become more concentrated and the chemistry would gradually become more efficient. So these things could evolve without any kind of replication.

FREEMAN DYSON, 1999

In the 1980s, following up the ideas generated during my sabbatical with Alec Bangham at Babraham, I began a research effort to understand how cellular life could have appeared on early Earth. It soon became clear that there were three major questions to be addressed: What kinds of lipidlike molecules were available to form membranous vesicles? How could something as large as a protein or nucleic acid be captured by a lipid vesicle to produce a protocell? And even if protocells could be produced, how could potential nutrients in the environment get across the membrane barrier to supply the trapped molecules so that they could grow and perhaps replicate themselves?

As described earlier, we already knew the answer to the first question. Amphiphilic molecules like fatty acids can assemble into membranous vesicles. Furthermore, they are present in carbonaceous meteorites, making it plausible that under certain environmental conditions, membranous vesicles were likely to be abundant in the prebiotic environment. But how could we take the next step, in which large molecules were captured inside the vesicles? I decided to test one possibility, which was to break down the permeability barrier by drying phospholipid vesicles mixed with large polymers such as proteins or nucleic acids. The idea was to simulate the kind of wet-dry cycle that would occur extensively at interfaces wherever atmospheric gases, water, and mineral surfaces met. Even though this seems like an obvious idea, no one had ever tried it before, partly because chemists interested in the origins of life preferred simpler reactions that can be studied in solution in a test tube. Also, most of the attention at that time focused on the concept of an "RNA World" in which life began as molecules of RNA that could serve both as catalysts and carriers of genetic information. This continues to be a fundamental concept in origins of life research and will be described in Chapter 13. But what it meant for me in the 1980s is that the origin of cellular life was being ignored, and the role that lipid membranes might play was wide open.

The basic experiments were easy to perform. We prepared liposomes in a test tube, added some nucleic acids or proteins, and dried the mixture with a gentle stream of nitrogen gas. We found that as they dried, the vesicles fused into extensive multilayered structures that trapped the proteins or nucleic acids between the layers. When water was added back to the dry film, the lipid layers formed vesicles again but now with up to half of the large molecules trapped inside.

This was an easy way to make protocells by simulating prebiotic wet-dry cycles, but I also wanted to visualize the process. Peter Armstrong was one of my colleagues in the Department of Zoology when I was at the University of California, Davis, and an expert microscopist. He agreed to lend a hand, so I made microscopic lipid vesicles by dispersing a few milligrams of phospholipid in 1 milliliter of water and then added the same amount of salmon sperm DNA to produce a viscous solution. I dried a small amount of the solution on a glass slide and then added a drop of water containing a dye called acridine orange that binds to DNA and produces a strong fluorescence. After I placed a very thin glass cover slip on the droplet, Pete put the slide onto the stage of his microscope and turned on the ultraviolet light to make the stained DNA glow.

"Wow!" he exclaimed and began to snap photos. One of these is shown in Figure 16. All around the edges of the dried lipid-DNA, where the water penetrated the dried mixture, microscopic structures containing fluorescent DNA were growing outward, taking the form of entwined tubules, spirals, and glowing vesicles. I had the answer I was seeking: It's easy to make protocells containing molecules even as large as DNA. All we need to do is to imagine that on prebiotic Earth, there would be continuous cycles of wetting and drying. If a membrane-forming lipid was present together with large polymeric molecules that had been synthesized under the same conditions, it seems inescapable that vast numbers of cell-like structures would be produced. Most of these would be inert, but a rare few might have just the right combination of molecules with catalytic and replicating properties that would permit them to grow and ultimately reproduce. This is the main concept that I want to put forward in this book, and later chapters will fill in the information required to establish tests of the idea.

WHY IS ENCAPSULATION IMPORTANT?

There are multiple reasons why encapsulated systems of molecules would be essential for life to begin. The most obvious is that encapsulation maintains a given set of molecules in one compartment, thereby allowing natural experiments to occur. To get a sense

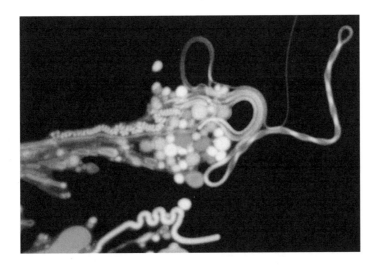

FIGURE 16
DNA encapsulated in lipid vesicles. The DNA has been stained with a fluorescent dye that marks the DNA encapsulated within the tubules and vesicles.

of why compartments are so important, imagine a chemistry laboratory in which the researcher is denied access to test tubes, beakers, and flasks, but instead must carry out all experiments by adding compounds to the ocean. In the absence of containers, how can an experiment be performed that requires specific mixtures of reactants? By analogy, only if self-assembled compartments were available on early Earth could specific mixtures of macromolecules be tested for their ability to interact with each other and with nutrients present in the environment.

The idea that membranes were somehow involved in the origin of life was first hinted at by John Burdon Sanderson Haldane, an eccentric but brilliant British scientist who wrote a short essay entitled "The Origin of Life," which was published in 1929. In the essay, Haldane speculated about conditions on early Earth. For instance, he proposed that there was little or no oxygen in the atmosphere, which instead would be composed largely of carbon dioxide. He imagined that there would be abundant organic compounds present, synthesized by the action of sunlight on mixtures of water, carbon dioxide, and ammonia, and he imagined that the first form of life would be something resembling a primitive version of a cell. Haldane wrote:

> The cell consists of numerous half-living chemical molecules suspended in water and enclosed in an oily film. When the whole sea was a vast chemical laboratory the conditions for the formation of such films must have been relatively favourable. . . . There must have been many failures, but the first successful cell had plenty of food, and an immense advantage over its competitors.

Haldane also stated that *"it is probable that all organisms now alive are descended from one ancestor..."* arguing that if there had been multiple origins, then there would be multiple kinds of chemistries, particularly related to the fact that all life today uses L-amino acids and D-sugars.

Haldane's thought process was amazingly prescient, but only in the last few years have researchers begun to seriously entertain the idea that life could begin as a replicating cell, rather than some kind of replicating molecule. The main reason is the general belief that lipid bilayers are virtually impermeable so that the membrane surrounding a primitive cell would make it impossible for it to have access to nutrients in the surrounding environment. This concern has been recently addressed by several research groups, so now I need to tell you about membrane permeability.

WHY ARE MEMBRANES BARRIERS TO FREE DIFFUSION?

The word "diffusion" has a specialized meaning in chemistry that needs to be explained. In a gas or liquid, molecules move around more or less freely in random directions, and this motion is called diffusion. It is easy to measure diffusion by separating two different gases or liquids with a barrier that is then removed. Over a period of seconds in the case of a gas, or days in the case of a liquid, the molecules no longer remain separated, but instead diffuse to produce a mixture. This process is called diffusion down a concentration gradient.

Living cells must maintain concentration gradients across their outer boundary membrane in order to stay alive, and this is why it is so important to understand how a membrane can act as a barrier to free diffusion of ions and nutrients. As described earlier, membranes are defined by lipid bilayers. It's an astonishing fact that this incredibly thin membrane, just two molecules thick, is essential for life to exist. The reason is that the oily interior of the bilayer largely excludes water, and this means that water-soluble substances such as glucose, amino acids, phosphate, and ions have very low permeabilities. Everyone has heard that "water and oil don't mix," but, in fact, water and oil do mix very slightly, except that the water dissolved in hydrocarbon chains is 10,000 times less concentrated than the surrounding water. This is what we mean when we say that lipid bilayers are permeability barriers. They greatly reduce the ability of water to get from one side to the other because water molecules don't readily dissolve in the oily interior of the bilayer. And yet, water molecules are small and lack electrical charge, so they are actually among the most permeable substances, along with gases like oxygen and carbon dioxide. When we measure the permeability of lipid bilayers to water-soluble solutes like amino acids and phosphate or simple ions such as sodium (Na^+) and potassium (K^+), the rates at which these substances get across lipid bilayers is a *billion* times slower than the rate at which water permeates the same barrier. The lipid bilayer barrier is essential to maintain the integrity of cells, but if a cell is to avoid starving to death, it must have access to external sources of nutrients. Modern cells solve this problem in a simple but elegant fashion. They synthesize proteins that insert into the lipid bilayer, which are specialized to transport a given solute across the bilayer barrier. These can be very complex molecular machines, such as the pump proteins that use energy to transport protons, calcium, sodium, and potassium ions across membranes. But other proteins just have an obvious pore through the center that allows ions to cross the bilayer. One of the simplest pore-forming peptides is gramicidin, an antibiotic synthesized by a soil bacterium called *Bacillus brevis*. Although antibiotics are now used to treat human disease, they originally evolved so that one species of microorganism could inhibit the growth of other species and thereby reduce competition for nutrients in the environment. Gramicidin, by the way, gets its name from Hans Christian Gram, who discovered in 1884 that certain kinds of bacteria could be identified by a special staining process he developed. Bacteria are now roughly divided into Gram-positive and Gram-negative varieties, depending on whether or not they can be stained. Gramicidin kills Gram-positive bacteria. In fact, the antibiotic ointments available for purchase at a drugstore often have gramicidin on the list of active ingredients.

The reason that pore-forming peptides like gramicidin are antibiotics is that they punch holes in the lipid bilayers of cell membranes and cause the membranes to become uncontrollably leaky. This is fatal because all living cells must maintain concentration gradients of ions in order to function. If an otherwise impermeable membrane suddenly becomes leaky when gramicidin penetrates the lipid bilayer, the ion gradients collapse and equilibrate. The result is that the cell can no longer produce energy and, like a car running out of fuel because of a leak in its gas tank, the cell ceases to function.

A computer model of the gramicidin channel is shown in Figure 17, and the central pore is clearly visible. Its diameter is barely sufficient to accommodate a few water molecules so that a single sodium or potassium ion traversing the pore literally pushes water ahead of it. The ion channels in cell membranes are a little larger. They are still pores,

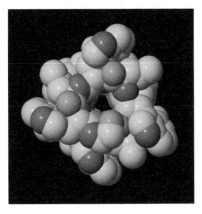
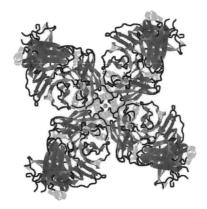

FIGURE 17

Gramicidin (left) is a small peptide composed of 15 amino acids that can assemble in lipid bilayers to produce a pore through a lipid bilayer. A similar top view of a potassium channel is shown on the right, in which the pore is produced by the arrangement of four protein chains. A single potassium ion is shown in the pore for scale.

but instead of a hole in a peptide like gramicidin, the pore is produced by several protein strands that crowd together in the lipid bilayer. Such channels are notoriously difficult to study. The structure of a bacterial potassium channel was only recently elucidated by a technique called X-ray crystallography, for which Roderick MacKinnon received the Nobel Prize in 2003.

WHAT NEEDS TO GET INTO AND OUT OF CELLS?

If we consider the first cellular forms of life, what would be required for them to grow? The basic list is relatively simple: Cells need a source of amphiphilic molecules, like fatty acids, for their membranes to grow, and they also need a source of monomers that can be incorporated into polymers by an energy-dependent process. Even though it is easy to consider this question conceptually, we don't yet know which of these components would be present on early Earth. But if we think about cellular life today, it seems reasonable to think that solute molecules such as amino acids, phosphate, nucleobases, and sugars are all possible candidates for what a primitive cell would need to grow, which brings us to the next question related to encapsulated systems of molecules. How could the first forms of cellular life have access to essential nutrients?

This seems like such a difficult question that it was probably responsible for the lack of interest in origins of life scenarios in which cells came first. But the apparent difficulty arises from two assumptions that are not entirely correct. The first is that lipid bilayers are impermeable. In fact, permeability is not a black-and-white issue, but instead bilayers have a continuous range of permeabilities. Small, neutral molecules like water, carbon dioxide, and oxygen can get through lipid bilayers with ease. For instance, blood travels through capillary beds in tissues in only a second or two, and oxygen carried by red blood cells diffuses out of the red cells and into tissue cells in that short time interval. There is more oxygen in the red cells than in the tissues because the mitochondria in tissue cells rapidly use up oxygen to produce energy. This produces a concentration gradient of oxygen across the red cell membrane, and molecules in solution always tend to diffuse from higher concentrations to lower concentrations at a certain rate. The result is that half the oxygen leaves the red cells in less than 1 second and diffuses into the cells of the tissue where it is rapidly used by mitochondria to synthesize ATP. Slightly larger neutral molecules (like urea and glycerol) diffuse across membranes in seconds or minutes. Only when we get to molecules the size of glucose and amino acids does the lipid bilayer begin to act as a significant barrier, and that's why cell membranes contain embedded transport proteins that help glucose, amino acids, and phosphate enter the cell.

The other assumption is that only biological phospholipids of a certain chain length can self-assemble into stable lipid bilayers. However, phospholipids are synthesized by enzymes which have evolved to use hydrocarbon chains 16 to 18 carbons long. This length is required to make the bilayers sufficiently impermeable so that ion concentration gradients can be maintained, but hydrocarbon chains that long were unlikely to be common on early Earth. If we use the organic compounds of the Murchison meteorite as a guide, the longest chains were probably 9 to 12 carbons in length. Can stable membranes assemble from amphiphilic compounds in this size range?

The surprising answer is *yes*. In 1978, Will Hargreaves, a graduate student in my lab, became interested in this question. Will was a remarkable young scientist with a rare passion to explore, and he soon found that an amazing variety of simple molecules could self-assemble into stable membranous vesicles, including fatty acids and hydrocarbon derivatives with phosphate or sulfate groups at one end. These could be as short as 10 carbons long, and Will discovered that membranes formed by mixtures of amphiphiles were usually more stable than the pure compounds. Once he called me over to the microscope and asked me to have a look. I did so, and saw the familiar microscopic vesicles swarming on the slide, magnified a few hundred times. "Those are pretty," I remarked. "What are they?" Will laughed. "Shampoo!" Will explained that shampoos were mixtures of amphiphilic molecules similar to those we were studying in the lab. Now when I take a shower, I see not only the foaming bubbles of the shampoo, but also in my mind's eye microscopic vesicles—the same kinds of self-assembled structures that gave rise to cellular life on early Earth.

We can now consider some typical nutrient solutes like amino acids and phosphate. Such molecules are ionized, which means that they cannot readily cross the permeability barrier of a lipid bilayer. If a primitive microorganism depended on passive transport of phosphate across a lipid bilayer composed of a typical phospholipid, it would require several years to accumulate enough phosphate to double its DNA content and divide into two daughter cells. In contrast, a modern bacterial cell with transport proteins in its membranes can reproduce in 20 minutes.

A possible answer to the permeability problem comes from Will's discovery that long chains are not required for stable membranes. Will's result was later followed up by Stefan Paula, another graduate student working in my lab, who found that simply shortening phospholipid chains from 18 to 14 carbons increases permeability to ions by a thousand-fold. The reason is that thinner membranes have increasing numbers of transient defects that open and close on nanosecond time scales so that ionic solutes can get from one side of the membrane to the other without dissolving in the oily interior phase of the bilayer. On early Earth, shorter hydrocarbon chains would have been much more common than longer chain amphiphiles, suggesting that the first cell membranes were sufficiently leaky so that ionic and polar nutrients could enter, while larger polymeric molecules were maintained in the encapsulated volume.

This idea was recently tested by Jack Szostak, who is a Howard Hughes Investigator in the Department of Molecular Biology, Massachusetts General Hospital. Jack is interested in the origin of cellular life and the possibility of producing synthetic life in the lab. Jack and his students published an important paper in *Nature* in 2008 with the following authors: Sheref Mansy, Jason Schrum, Mathangi Krishnamurthy, Sylvia Tobé, Douglas Treco, and Jack Szostak. This might be a good place to mention that publications in major journals like *Nature* often have multiple authors in order to give everybody proper credit for their contributions. The first authors, in this case Mansy and Schrum, are typically graduate students or post-doctoral associates who did most of the actual bench work and probably wrote the first draft of the paper. The last author is usually the principal investigator who directs the research and was awarded a grant from a national funding agency to support the work, in this case NASA and the National Science Foundation. The authors listed in between contribute to the research in a variety of other ways.

The 2008 paper addressed the question of how growing systems of polymers in a primitive cell could have access to nutrients present in the external environment. The team answered this question by fabricating compartments not from modern phospholipids, which have evolved to be virtually impermeable, but from mixtures of simpler molecules like fatty acids, fatty alcohols, and monoglycerides. Mansy investigated the permeability of such vesicles and optimized the mixture so that small solute molecules could pass through the membrane, but larger polymers could not. Mansy and Schrum then captured a synthetic DNA molecule that could act both as a primer and template for base-specific elongation if it had access to an activated substrate. The substrates, analogous to nutrients, were added to the external medium as activated nucleotides. The template strand was a string of cytosine bases, so the nucleotides added were the complementary guanosine nucleotides. The reaction was monitored for several hours to see what happened.

If the vesicles were composed of phospholipids, there was no reaction—as expected— because the nucleotides could not permeate the lipid bilayers. But the vesicles composed of mixtures of short-chain amphiphilic molecules showed a remarkable growth of the DNA molecules as guanosine nucleotides were added one by one to the primer strand, with the cytosine string acting as the template.

Previous studies have demonstrated a variety of internalized polymerization reactions, but the Szostak group put it all together into a working system with a membrane that permits nutrient transport. One of the most important implications of their research is that a primitive version of a cell can be heterotrophic, using nutrients available in the environment, such as sugars and amino acids, directly as monomers for growth processes. This is in contrast to more complex forms of life that are autotrophic, in which metabolic pathways can synthesize monomers like amino acids from carbon dioxide and nitrogen using light as an energy source. This is a much more complex task, so it seems reasonable to assume that the first cells were heterotrophic and used nutrients available in the prebiotic environment.

ENCAPSULATED SYSTEMS CAN CAPTURE ENERGY

Because they are such central molecules in life processes, it is common knowledge that nucleic acids and proteins are essential for life and that genes are encoded by the sequence of bases in DNA. However, an equally fundamental fact is that living cells today are absolutely dependent on concentration gradients of ions. Even the simplest bacterial cells use energy to transport protons, sodium, and potassium ions across their boundary membranes, and the resulting ion gradients are an energy source for ATP synthesis, nutrient transport, and motility. All of these functions depend on the relative impermeability of lipid bilayers to ions. If the membranes were too leaky, the cellular life that dominates the biosphere today would not be possible.

Now we come to a conundrum. Life today depends on ion concentration gradients, and we imagine that the first forms of life probably used concentration gradients in some way. And yet, the first cell membranes were necessarily leaky in order for nutrients to enter the cell to supply the necessary ingredients for growth by polymerization. No one has yet initiated the research effort that will be necessary to resolve this conundrum, but it is possible to offer some ideas about how to approach the problem. The first thing is to question the assumption that transport proteins were not available. We assume that the first living cells were able to synthesize polymers capable of catalysis and replication, so why not go one step further and propose that some of the polymers being synthesized happened to be pore-forming molecules that allowed specific ions and nutrients to enter the cell? Only those cells that had this capability would be able to survive and grow.

Ann Oliver, who carried out her doctoral research in my lab, was among the first to test this idea experimentally. Ann prepared lipid bilayer membranes and exposed them to simple peptides consisting only of the amino acids alanine or leucine. These are hydrophobic amino acids, so the peptide chains could become embedded in the hydrophobic phase of the lipid bilayers. Ann found that the peptides spontaneously organized themselves in the bilayer to produce channel-like defects that could conduct protons (but not sodium or potassium) across the bilayer. This result makes it plausible that similar peptides produced channels that primitive cells could use to transport specific ions or nutrients across membranes.

It may be significant that the peptides allowed a specific conductance for protons. As discussed in Chapter 6 on prebiotic sources of energy, a variety of chemical and photochemical reactions feed electrons into a reaction pathway called electron transport. During electron transport, a second tightly coupled reaction drives protons across membranes to produce a proton gradient, and the leak of protons back through a specific enzyme called ATP synthase produces ATP. Almost no one except biochemists realizes that proton gradients are the power source for all life on Earth, so I hope that readers will add that fact to their personal collection of essential knowledge.

An important point to understand is that the proton gradients essential for life are produced and maintained by living cells despite a surprisingly rapid leakage of protons through the membrane. Years ago, Wylie Nichols carried out his doctoral research in my lab and discovered that protons have a special conductive pathway which allows them to get across lipid bilayers thousands of times faster than other cations. For instance, if we isolate the proton-pumping membranes from a green plant like spinach and expose them to light, a proton gradient is produced in a few seconds; but if we then turn off the light, the proton gradient decays in less than a minute, while a similar gradient of sodium or potassium ions would take hours to leak out. The point is that even modern cells have not found a way to produce a membrane that is impermeable to protons, but instead they simply pump protons fast enough to produce a gradient in spite of the leak. I think that a very fruitful area of future research will be to search for chemical mechanisms by which primitive cells could produce proton gradients across similarly leaky membranes. If such a mechanism could be found, it would represent a fundamental breakthrough in our understanding of the origin of life because we would finally have a link between a universal energy-transducing process today and a primitive version that could supply an energy source for early life.

HOW CAN ENCAPSULATED SYSTEMS GROW?

Another cutting edge of research on the origins of cellular life is to gain some understanding of how primitive cellular systems could grow and then divide. Again, there are no easy answers, but there are some very exciting leads. Pier Luigi Luisi was one of the first researchers to recognize the significance of this question. Luigi is an unusual scientist who has as much appreciation of the arts as the sciences. He is the organizer of an annual gathering at Cortona, Italy, where scientists and artists mingle in the unique atmosphere of an abbey that has been turned into a conference center. I was among the participants some years ago, and after the usual tedium of international travel and then a train ride to Cortona, I fell into an exhausted sleep in a little room formerly occupied by a monk, but it was now quite comfortable. When I awoke the next morning with the Tuscany dawn, someone in the courtyard below was greeting the sunrise by softly playing an achingly sweet Debussy melody on her violin. I knew then that I was not at a typical scientific conference. Over the next few days, it was a remarkable experience to mingle with musicians, sculptors, artists, scientists, and priests in an atmosphere of mutual respect and friendship.

In the 1990s, Luigi and his research group in Zurich began to work on vesicle systems composed of oleic acid, an 18-carbon fatty acid which forms particularly stable membranes at pH 8.5. The group wanted to get the vesicles to grow and came up with an elegant idea: Why not "feed" the vesicles by allowing a precursor to produce oleic acid? The precursor is oleic anhydride, which is composed of two oleic acid molecules held together by an anhydride chemical bond. The word "anhydride" has a similar sense as "anhydrous" meaning without water. Anhydrides are formed by removing a water molecule from between two acidic groups, such as the carboxyl groups of the oleic acid. Anhydrides have a tendency to break apart in the presence of water because water molecules spontaneously add back to the anhydride bond—a process called hydrolysis. The result is that two oleic acid molecules are produced from each molecule of the anhydride.

In a paper published in 1994, Luigi's group demonstrated that when oleoyl anhydride was added to a water solution buffered at pH 8.5, oily droplets of the anhydride at first floated on the surface but over time, hydrolyzed and spontaneously formed vesicles. Remarkably, if the anhydride was added to a solution containing previously formed vesicles, the reaction proceeded much faster, as though the existing vesicles catalyzed the formation of new vesicles. These results made it clear that vesicles can grow by simple addition of fatty acids in the medium to existing membranes.

Luigi's pioneering work inspired a more recent study by Martin Hanzyck and Ellen Chen, who were graduate students working in Jack Szostak's laboratory. Their aim was not just for the vesicles to grow, but also to divide and then grow further. Martin used a fatty acid with a hydrocarbon chain having 14 carbons rather than the longer 18-carbon chain of oleic acid. After the vesicles began to grow, he passed them through a filter. As the larger vesicles entered the pores in the filter, they broke up into smaller vesicles that could then start growing again. Significantly, if RNA was encapsulated in the larger vesicles, it was not lost when the vesicles were filtered to produce smaller vesicles. This remarkable series of experiments clearly demonstrated the relative simplicity by which a complex system of lipid and genetic material can undergo a primitive form of growth and division.

CONNECTIONS

The connection between the origin of life and cellular compartments is relatively recent, mostly because workers in this field prefer to study reactions in solutions. I like to remind my chemist colleagues that if they think compartments are not essential, they should try running their reactions in the open ocean. The fact is that the test tubes, flasks, and beakers that chemists use to contain their reactions are, in essence, just big compartments. Furthermore, they are not impermeable, but instead have a large opening at the top that is analogous to the pore in a cell membrane through which reactants (nutrients) enter and products leave. It would be impossible for chemists and biochemists to perform their experiments in the absence of encapsulated systems called test tubes. By analogy, it seems reasonable to think that the origin of life also required encapsulation.

A number of essential properties emerge when compartments are added to the scenarios proposed for the origin of life. Most obvious is that compartments with boundaries composed of lipid bilayers keep molecular systems together so that their components can interact. However, the boundary structures must be sufficiently permeable so that the encapsulated systems can have access to small nutrient molecules in the environment, yet they must be sufficiently impermeable to keep larger molecules in one place. Recent experiments show that membrane boundaries composed of simple amphiphilic molecules have this property. Furthermore, such boundary structures can grow by addition of amphiphilic molecules from the environment and can even divide if they are subjected to modest shear forces produced by turbulence. Less obvious is that boundary membranes composed of lipid bilayers provide a non-polar environment that is necessary for certain reactions, particularly those related to capturing energy required for life processes. For instance, capture and transduction of light energy by pigments in plant cell membranes provides most of the energy for today's biosphere. A final point about encapsulation is that only in this way can specific molecular systems be produced that initiate biological evolution.

The last connection to be made is that the origin of life can be considered to have been an immense exercise in combinatorial chemistry. The robotic devices that run thousands of parallel experiments in a day are marvelous examples of the power of biotechnology, but their capacity is less impressive when they are compared to a test tube of phospholipids dispersed as vesicles, which potentially represent trillions of parallel experiments. Given an entire planet, millions of tons of organic compounds, and half a billion years, it seems inevitable that primitive cellular life will arise through a natural version of combinatorial chemistry. The goal now is to discover the compounds and conditions that will let us repeat the process in the laboratory.

9

ACHIEVING COMPLEXITY

There are living systems: there is no living 'matter'. No substance, no single
molecule, extracted and isolated from a living being possess, of its own, the
aforementioned paradoxical properties. They are present in living systems only;
that is to say, nowhere below the level of the cell.

JACQUES MONOD, 1967

François Jacob and Jacques Monod were first to demonstrate that genes are tightly regulated and can be switched on and off as required in order to respond to changes in the
environment. Jacob and Monod invented a new name for the switch, calling it an operon
because, in a sense, it operates the machinery of the cell. For their discovery, Jacob and
Monod shared the Nobel Prize in 1965.

Monod's quote above is perfect for introducing this chapter. It makes a point that is
central to the question of how life began, which is that the first forms of life were necessarily systems of interacting molecules. Although replicating molecules like nucleic acids
and catalytic molecules like enzymes are essential for life, by themselves they are not
alive. Only when they are incorporated into a system of other molecules can that system
take on the properties of life. The system must encapsulate its component molecules in
a compartment of some sort—the membranous boundary of all living cells. This chapter
describes several of the most important systems of life and how they are regulated, and it
then asks how the first systems of interacting molecules could emerge when life began.

The fact is that even the simplest living cells—bacteria—contain thousands of different molecules organized into intricate, regulated networks. New words like "genomes,"
"proteomes," "transcriptomes," and "metabolomes" are being coined that show where
our understanding of living systems is headed. The suffix "ome" has the sense of "whole"
or "total." Pioneering researchers are now mapping out the interactions between the
major protein components of living cells, and the resulting maps are called (of course)
interactomes!

The word "system" is derived from a Greek word having to do with being orderly, organized, and connected. Today the word is widely applied to everything from political systems to solar systems, but here we will use it in a specific biological sense: *In living cells, systems are complex sets of molecular components that interact in order to carry out a specific function and are regulated by a variety of control mechanisms.*

There are four general systems that form the basis of all life today: a system of enzymes that catalyze and guide metabolic reactions, a second system of enzymes and membranes that produce energy for the cell, a third system of enzymes and ribosomes that synthesize proteins using the genetic information in nucleic acids, and a fourth system of enzymes that replicate the nucleic acids so that genetic information can be passed to the next generation. There are many other systems, of course, such as those responsible for transport of nutrients across membranes, cell division, sensory response, and motility, but the four outlined above are probably the most fundamental to the definition of life. How systems could spontaneously arise is a question that virtually no one has yet addressed in origins of life research, so it is wide open for future investigations.

SYSTEMS AND BIOCOMPLEXITY

As one example of a regulated system, we can consider glycolysis, which was described briefly in Chapter 6 on energy. Glycolysis is a series of enzyme-catalyzed reactions that extract energy from glucose and make the energy available to the cell, along with the products of the reactions. For instance, when yeast cells grow in grape juice during fermentation at a winery, no oxygen is involved; only glycolytic reactions occur, which the yeast uses as a source of energy. Intermediate products of glycolysis are siphoned off to make other necessary biochemical components, such as amino acids, while carbon dioxide and ethanol are end products of this particular system.

The first step in glycolysis occurs when an enzyme called hexokinase transfers a phosphate group from ATP to glucose. If we made a mixture of equal amounts of ATP and glucose, added a little hexokinase, and then analyzed the mixture a few minutes later, we would find that most of the glucose would be in the form of glucose phosphate, which is a glucose molecule with a phosphate group attached:

$$\text{Glucose} + \text{ATP} \{\Rightarrow\} \text{glucose-6-phosphate} + \text{ADP}$$

The phosphate comes from ATP (adenosine triphosphate), so ATP is used up during the reaction and only ADP (adenosine diphosphate) remains. The hexokinase by itself would not be considered to be a system because it catalyzes just one reaction. But when hexokinase is involved in the glycolytic pathway that metabolizes glucose, the series of enzymes is considered to be a system.

Glycolysis does not proceed at full speed all the time, but instead is regulated by feedback control. Understanding feedback is important if we are to understand the metabolic reactions that characterize a living cell because virtually every biochemical system in a cell is regulated by feedback. The hexokinase enzyme described above is controlled by a feedback mechanism called product inhibition. The product of the reaction is glucose-6-phosphate (G-6-P), which can also bind to the active site on the hexokinase. This prevents glucose from binding and thereby inhibits enzyme activity if too much G-6-P is being produced.

To sum up, the regulated system of glycolytic enzymes is more complex than any single component, so complexity is related to the idea of systems. Furthermore, feedback interactions among the components somehow control the overall function of the system. Before describing examples of biological systems related to the origin of life, we should get a sense of what we mean by complexity, and more specifically biocomplexity. Then we can ask how complex living systems might emerge from disorganized mixtures of organic components.

BIOCOMPLEXITY IN LIVING SYSTEMS

The concepts of systems and biocomplexity go hand in hand. Systems are real and can be defined, but as you will see, it is more difficult to measure complexity. However, it is still a useful word, particularly in describing what happened during the four billion years of evolution from the first forms of life to the biosphere we now inhabit. The concept of complexity can seem pretty abstract at first because we don't have a way to attach quantitative meaning to the word. For instance, there is no equation that allows us to calculate whether the brain of a bird is more complex than the brain of a bee, even though it might seem obvious. On the other hand, perhaps we can use a comparison of calculators and computers as a guide for estimating complexity. To describe how much more complex a personal computer is than a calculator, we count and compare the number of components in each device, then we compare the number of possible interactions between components (the structure of the logic circuits) and how fast the functional interactions can occur. Maybe we can do something similar with a living system, because we can, in fact, count the number of different kinds of molecules in a living cell, the number of metabolic pathways, and the regulatory interactions between components of the pathways. In other words, we can understand life in terms of the relative complexity of the systems that permit life to exist. We can then ask a deep question: What is the simplest system of molecules that can be called alive? If we can answer that question, it will provide insight into the systems of molecules that self-assembled in the prebiotic environment to give rise to life.

We can get an intuitive sense of biocomplexity by considering examples of living systems. For instance, a bacterial cell seems more complex than the nutrient medium in which it is growing, even though the cell and its growth medium are composed of exactly the same set of atoms. It follows that one parameter of biocomplexity has to do with the degree of structural organization of molecular components, which are organized in a bacterial cell but disorganized in the growth medium. A living cell seems more complex than a dead cell, so biocomplexity is also related to the organized network of dynamic functions that occur within the structure of a cell. A multicellular organism is more complex than a single-celled organism, which means that complexity is related to the number of components in the system. Finally, a multicellular organism with a nervous system is clearly more complex than an organism lacking one, suggesting that the network of interactions among components also contributes to biocomplexity.

These all seem like reasonable statements, but how can we establish even the simplest semi-quantitative description of what we mean by "more complex than"? To narrow the question further, we can try out some quantitative terms and test them against our intuition. Is a bacterial cell twice as complex as its growth medium? That seems much too little. Ten times? A hundred times? That's still too little, but getting closer. At the other end of the complexity scale, is a bacterium infinitely complex? This seems too much of a leap. A bacterial cell is a finite structure, so the answer is somewhere between twice as complex and infinitely more complex.

STRUCTURAL COMPLEXITY

We can begin by considering increments in complexity related to the molecular components that gave rise to the origin of life.

1. The sterile surface of early Earth became more complex with the addition of organic compounds, either by synthesis or by delivery during accretion, as described in Chapter 4.

2. In turn, the mixture of organic solutes became more complex over time as organic molecules self-assembled into molecular aggregates. Examples include the chemical synthesis of random polymers from suitable monomers such as amino acids, and the assembly of membranous vesicles from amphiphilic molecules.

3. The nascent biosphere became much more complex when one or more of the self-assembled structures happened to have properties that allowed it to use energy to accumulate simpler molecules from the environment and assemble them into reproductions of the original structure.

4. After life began, biocomplexity increased further as macromolecular structures became organized into systems within the cellular unit of life in order to catalyze metabolic pathways and to transmit information from one kind of molecule to another. Examples include the genetic code, replication, transcription, and translation.

Biological complexity did more than simply emerge. The first living cells also had the capacity to evolve. Biological evolution began when prokaryotic cellular life filled niches and began to compete for energy and nutrients. Because there were variations in composition and structure from one cell to the next, individual cells differed in their ability to grow and reproduce in a given niche: Some thrived, others died. The result of the competition was that different kinds of microorganisms appeared, each specialized to fill a certain environment on early Earth. We would now call this process speciation, and complexity further increased with the number of species competing for the niches.

Significantly, the converse is also true. Evolution is driven not only by competition and "survival of the fittest" but also by interactions within or between species such that populations of organisms have a better chance of survival and reproduction than an individual organism. For instance, bacteria today only transiently live as single cells. As they multiply, they form structures called microbial mats that offer protection and mutual assistance to the individual cells, and the mats represent yet another increment in biocomplexity.

To complete this story of ever-increasing complexity, at some point, microbial populations reached the carrying capacity of the environment, and the level of biocomplexity stabilized for a period of approximately two billion years. Around 1.8 billion years ago, several varieties of microbial life underwent symbiotic mergers to produce the first eukaryotic cells, some of which had autotrophic lifestyles in which photosynthesis played a major role, while others were heterotrophic, using the energy and nutrients produced by the autotrophs. Another billion years passed while oxygen produced by photosynthesis slowly accumulated in the atmosphere. Populations of eukaryotic cells, given a new energy source of respiration, began to assemble into multicellular organisms. Certain cells in multicellular organisms differentiated into tissues with specialized functions, particularly the cell-to-cell communication pathway called a nervous system. This allowed one cell to affect large numbers of other cells by efficiently transmitting a kind of digitized signal. Complexity increased further as the size and variety of multicellular organisms grew following the Cambrian explosion. Over the next half-billion years, marine organisms first occupied the intertidal zones between seas and land and then ultimately found ways to live beyond the marine environment as plants, animals, fungi, and microbial life—all interacting within the ecosystems that characterize the biosphere today.

FUNCTIONAL COMPLEXITY

The second aspect of biocomplexity is functional complexity, and metabolism is an important example of functional complexity of molecules related to the origin of life. There is no simple way to illustrate the reality of metabolic pathways in a cell. The problem is that all of the reacting molecules are mixed in the cytoplasm and randomly diffusing. The reactions are occurring simultaneously, and they are controlled so that a given reaction cannot go too fast or too slow. Furthermore, most of the reactions are reversible and can proceed in either direction. The reason there is an overall direction to metabolic reactions is that they proceed from higher to lower energy levels (they run "downhill") and carbon dioxide—the end product—escapes from the cell into the surrounding environment. If it did not, metabolism would quickly grind to a halt and the cell would die.

To get a sense of how a metabolic system functions, imagine that we inject a pulse of labeled compounds into a cell and then follow the sequence of changes that occurs over time. Over the past century, biochemists have laboriously performed these kinds of experiments and we now have a pretty good idea of the major metabolic pathways that cells use. A simplified diagram is shown in Figure 18, in which we begin with three major components of all life, lipids (fats), carbohydrates (starch), and proteins. The passage of time is indicated from left to right. If you actually did the experiment, the changes would occur in only a few seconds. The energy content of the molecules is indicated from top to bottom, showing how the reacting molecules tend to lose energy as they go through the reactions. The molecules themselves are shown as light clouds against a darker background to indicate that we are not watching the changes happening to a single molecule, but instead to large numbers of molecules going through a sequence of reactions.

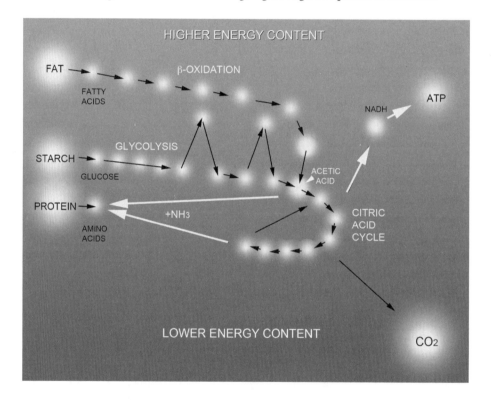

FIGURE 18

The map of metabolic pathways is an example of the complexity of living cells. The white puffs represent populations of biochemical compounds called metabolites, and the black arrows between puffs are reactions catalyzed by specific enzymes. The larger white arrows indicate chains of complex reactions with many enzyme-catalyzed steps involved. For instance, glycolysis starts with glucose at the left and ends in acetic acid, which then enters the citric acid cycle in mitochondria where it is oxidized to carbon dioxide, releasing most of the energy available in the original glucose. Fat has about twice the energy content, and its energy is released by stepwise breakdown to acetic acid. Significantly, some of the intermediates of glycolysis and the citric acid cycle can be metabolized to amino acids by addition of ammonia (NH_3).

The downhill reactions begin when fat, starch, and proteins are broken down into their simpler components: fatty acids, glucose, and amino acids. Note that fat has roughly twice the energy content of starch and protein. This is because the chemical energy is stored in the electrons between carbon and hydrogen. Most of the carbon in fat is present as long chains of methylene ($-CH_2-$) groups in which two hydrogens are bonded to each carbon atom, while in carbohydrates most of the carbons are in a partially oxidized form in which a hydroxl group has replaced one of the hydrogens ($-HCOH-$). In carbon dioxide, the end product of energy metabolism, the carbon is fully oxidized ($O=C=O$) and has no remaining chemical energy. An example of a subsystem within the map is the pathway of glycolysis, which starts at the top with glucose and leads down to acetic acid just above the circular ring of reactions. The circle at the bottom is called the citric acid cycle, or sometimes the Krebs cycle in honor of Hans Krebs who published his discovery of the cycle in 1937 and was awarded the Nobel Prize in 1953.

The most important point to understand is that the major metabolic pathways are connected. One such connection point is acetic acid, in which the metabolism of fatty acids, glucose, and amino acids all intersect. Pyruvate is a connection point between glycolysis and amino acid metabolism because ammonia (NH_3) can be added to pyruvate to produce the amino acid alanine. The citric acid cycle is a connection between glycolysis and energy production because the cycle produces a high-energy compound called NADH that supplies electrons for ATP synthesis in mitochondria. The citric acid cycle also represents a kind of "drain" for all of metabolism because this is where most of the carbon dioxide is produced that is released into the environment. If the drain gets plugged up, the entire flow of energy grinds to a halt and the organism dies.

A map like that shown in Figure 18 is very schematic and should not be considered to represent what actually goes on in a cell. Because substrate molecules are free to diffuse in all directions throughout the cell, reactions do not take place in straight lines. Instead, they occur only by chance when a specific molecule happens to collide with and bind to the active site of an enzyme that catalyzes a reaction. Often two different molecules must bind to two different sites on a given enzyme, such as glucose and ATP on a hexokinase enzyme. Only then can a phosphate be transferred from ATP to glucose to produce glucose-6-phosphate. Finally, most of the reactions are reversible, which means that we can start with acetic acid and from that synthesize fatty acids, glucose, and amino acids.

Glycolysis and the citric acid cycle are important to the origin of life for several reasons:

- The reactions are relatively simple compared to the rest of metabolism and may provide insight into the origin of metabolic pathways. For instance, there is no nitrogen in the biochemical components of glycolysis and the citric acid cycle. Instead, amino acids are synthesized from intermediates by adding nitrogen in the form of NH_3, or ammonia: serine and glycine are synthesized from phosphoglyceric acid, alanine and valine from pyruvic acid, aspartic acid from oxaloacetic acid, and glutamic acid from alpha-keto glutaric acid. It is as though a simpler version of the glycolysis and the citric acid cycle reactions may have existed before the origin of life, and those simpler versions were then adopted as a backbone for other metabolic pathways.

- Glycolysis is a source of energy that does not require oxygen. Therefore, it is an example of the way that energy can be extracted from a series of linked chemical reactions that could occur on prebiotic Earth, before oxygen was available.

- Under certain conditions the citric acid cycle can run in reverse so that carbon dioxide (CO_2) is taken up rather then released. The primary function of photosynthesis in the biosphere today is to capture carbon dioxide in the atmosphere and make the carbon atoms available for life processes. In a very real sense, the leaves, stems, and trunks of every plant are composed of carbon atoms that were once present in the atmosphere as gaseous CO_2. Because the nutrients used by all animals ultimately come from plants, those same carbon atoms now compose the cells and tissues of your body. Once early life got started by assembling and then using organic molecules that were already present in the prebiotic environment, microorganisms needed to capture carbon dioxide from the atmosphere; otherwise all available nutrients would soon be used up. A reversible citric acid cycle suggests one way in which carbon dioxide could be incorporated into living systems in the absence of photosynthesis.

- Glycolysis is regulated at several points, as will be described in the next section. In any dynamic system, including the primitive metabolism of the first forms of life, there must be regulatory mechanisms; otherwise one of the reactions will far outpace the rest, and the system is disrupted either by starvation or glut.

CONTROL PROCESSES IN COMPLEX SYSTEMS

One of the essential properties of life is that the rates at which enzymatically catalyzed processes take place are regulated. The result is called homeostasis, which means that the systems of life are maintained within certain set limits. The regulation is mostly done by feedback loops, a term that can be confusing until we realize that feedback also controls a number of processes that are part of our everyday lives. One example is the way that furnaces are controlled by a thermostat, and the loop here is defined by a sensor, a signal, and an effector, with the sensor having an output that loops back to control the effector. The sensor in this case is a thermometer in the thermostat that measures the temperature of the house. The thermostat generates an electrical signal that closes a switch when the temperature falls below a certain set limit and opens the switch when the temperature is above that limit. When the switch is closed, the furnace turns on to heat the house, and the switch opens when the desired temperature is reached. This means that the actual temperature oscillates a few degrees above and below the desired temperature, which is characteristic of feedback loops. Another familiar, but more complex feedback loop is the cruise control that maintains the forward motion of an automobile at a certain speed.

Both thermostats and cruise controls are simple on-off mechanisms. The furnace actively heats the house but does not cool it, and the cruise control actively accelerates but does not actively brake your car. In complex living systems, processes typically have both an accelerator and a brake to provide a more precise control. For example, the level of glucose circulating in the blood is regulated by two hormones called insulin and glucagon. Insulin lowers blood sugar by increasing the rate at which it is transported into cells, while glucagon raises blood sugar by increasing the rate at which glucose is released from a storage form in the liver called glycogen. But at the cellular level, most of the feedback loops involve just braking, in which one or more enzymes in a system are inhibited by a product. For example, three of the enzymes involved in glycolysis use ATP, and the reactions they catalyze are largely irreversible, which makes them good candidates for regulatory control.

BIOCOMPLEXITY OF INTERACTING MOLECULAR SYSTEMS: INTERACTOMES

Earlier in this chapter, we compared the relative complexity of calculators and computers in terms of the number of components and the number of functional interactions. Now we can use this analogy to get a sense of the relative structural and functional complexity of the systems of molecules within a living cell. The number and kind of interacting structural-functional units can be reasonably well estimated for a bacterial cell, and Table 3 shows the actual count in *E. coli*.

This is an astonishing example of the power of molecular biology today. My personal computer has only 1,000 or so components in a volume of perhaps 5 liters, but life has

TABLE 3   Number of Molecules in a Single *E. coli* Cell

Molecular Components	Number of Molecules
Kinds of proteins	1,850 different proteins, mostly enzymes
Total number of proteins	2.36 million
RNA in ribosomes	18,700
Transfer RNA	205,000
Messenger RNA	Variable, depending on growth cycle
DNA	1 circular double helix
Lipid	22 million (mostly in cell membrane)
Lipopolysaccharide	1.2 million
Glycogen	4,360 (energy storage for the cell)

found a way to pack millions of components into a volume of a few cubic micrometers! Not only that, but a cell can also grow and reproduce itself, something that a robotic device with a computer brain can only do in science fiction.

Now that we have an estimate of the number of components, let's consider the number of interactions between components. New techniques have made it possible to establish which proteins interact and the number of interactions. This has given rise to a concept called "interactomes," which are defined in terms of the total number of protein species in a living cell, each of which is functionally linked by one or more interactions with other proteins in the cell. Interactomes have now been reported for the bacteria called E. coli, the yeast S. cervacia, the nematode C. elegans, the fruit fly D. melanogaster, and even Homo sapiens.

Examples of interactomes can be found in the sources for this chapter listed at the end of this book. An interactome does not exist as such in a cell, but instead is a map that describes which proteins have a functional interaction and which do not. The resulting maps look like furry balls composed of thousands of small circles connected by lines. Each circle represents a protein, and the lines indicate one or more interactions between a given protein and other proteins. If a circle has only a single line, it means that only a single interaction has been established. Other balls might be at the center of multiple lines because they are interacting with a dozen or more other proteins in the cell. Whenever a unit within a system has more than one interaction with other units, the complexity of the system increases exponentially.

The concepts of biocomplexity, regulated systems, and interactomes bring us to the most important question in this chapter: If regulated systems of interacting molecules are so important to all life today, were they also essential for the origin of life? And if so, can we conceive of a process by which they might arise?

THE ORIGIN OF SYSTEMS AND THE MINIMAL CELL

This chapter has made the point that the origin of life could also be considered to be the origin of molecular systems that had certain specific properties. On prebiotic Earth, countless numbers of natural experiments were going on, representing a massive process of combinatorial chemistry, and life began when one of the experiments—a membranous compartment containing a specific mix of macromolecules—began to grow by energy-driven polymerization and then to replicate the macromolecules. Based on what we know about biological systems today, I think it may be possible to conceive of a hypothetical first living system, and perhaps even discover a way to assemble such a system in the laboratory.

First, we need to define the organic compounds that are assumed to be present as nutrients for the system. These include a mixture of amino acids, a second mixture of compounds resembling nucleotides, and third a mixture of amphiphiles capable of self-assembling into a membranous compartment. Using energy available in the environment, a catalytic subsystem chemically changes nutrients into usable monomers and then activates the monomers so that they are able to polymerize by a catalyzed reaction. Two kinds of polymers are required. One species catalyzes the polymerization of the activated monomers into a second species of macromolecule that is also a catalyst. A second species of macromolecule catalyzes the replication of the first species. These catalysts are maintained in a membranous compartment that grows by addition of amphiphilic molecules. The boundary of the compartment allows small monomers into the internal volume, but it captures the polymers.

Now we can define the control points. There must be feedback control between the polymerization reaction and the replication reaction; otherwise too much of one or the other macromolecule will be synthesized. A second feedback must regulate growth of the membrane and growth of the polymers, and a third feedback must regulate the synthesis of activated monomers. No one has yet attempted to develop an experimental system that incorporates all of the above components and controls, so we can only speculate on how control systems might have developed in early forms of life. There is one obvious point in the network that offers a place to start. Nothing can happen unless small nutrient molecules can get across the membrane boundary, so the rate at which this happens will clearly control the overall process of growth. I propose that the first control system in the origin of life involved an interaction of internal macromolecules with the membrane boundary as the signal and that the effector of the feedback loop was the permeability of the bilayer to small molecules. As internal macromolecules were synthesized during growth, the internal concentration of small monomeric molecules would be used up and growth would slow. But if the macromolecules disturbed the bilayer in such a way that permeability was increased, this would allow more small molecules to enter and support further growth, representing a positive feedback loop.

Is there any experimental evidence that supports these conjectures? Michael Yarus and his research group at the University of Colorado have in fact demonstrated that RNA can be selected to interact with lipid bilayers. The resulting RNA species disturbs the bilayer barrier in such a way that it becomes more permeable to ions and even exhibits channel-like conductance. Furthermore, it has been known for years that an undisturbed lipid bilayer is in an equilibrium state that resists the insertion of additional lipid molecules, but if it is exposed to a pressure gradient, extra lipid molecules readily add, and the bilayer grows. This, by the way, is exactly what happens when we blow a soap bubble, in which the pressure differential permits extra soap molecules to enter the bubble membrane as it grows. It follows that the growth of the boundary membrane necessary to accommodate the growth of internal polymers could be regulated by the disturbance of the bilayer barrier by the macromolecules. The disturbance allows more amphiphilic molecules to get into the bilayer so that it will also grow.

STRESSES IMPINGING ON THE FIRST LIVING SYSTEMS

Once systems were in place that acted to produce homeostasis, the earliest life discovered ways to respond to stresses, both physical and chemical. A stress is generally defined as an excursion in environmental conditions that tends to disturb otherwise homeostatic systems. The physical stresses impinging on early life are obvious: temperature changes (either too hot or too cold), osmotic stress due to changes in concentration of solutes outside the cell, radiation in the form of ultraviolet light, and mechanical stresses such as shear forces resulting from agitation of aqueous environments. The chemical stresses would be represented by the constant battle against denaturation and hydrolysis of polymers, the effects of pH and divalent cations on those same polymers, periodic lack of nutrients leading to starvation, and a variety of toxic substances to which the cells might be exposed due to variations in the chemical composition of the environment. Life today has evolved superb defensive measures against most if not all of these stresses, and we well understand the limits of life in its ability to respond to stress. However, the first forms of life could not have had much in the way of defensive systems. This means that the first living systems could have been very delicate, requiring a benign environment to exist. Alternatively, the first life might have consisted of simple, highly robust systems of molecules that were relatively immune to environmental stresses. We don't know which is more plausible yet. We can only make guesses and then test our ideas. I would favor a relatively benign environment for the origin of life, in order to minimize the dispersive effects of temperature, pH, and ionic conditions on self-assembly processes. Soap bubbles require just such an environment to exist. The assembly of soap molecules into bubbles is a delicate process that cannot occur at elevated temperatures, acidic pH ranges, high divalent cation concentrations, or strong shear forces. The microscopic versions we call vesicles are less affected by temperature and mechanical forces, but they resemble soap bubbles in that they can only exist in narrowly defined conditions of pH and ionic composition.

CONNECTIONS

The point made in this chapter is that the origin of life is best understood as occurring within a hierarchy of increasingly complex systems of molecules governed by chemical and physical laws. There must have been a process by which large polymeric molecules were synthesized, and a rare few of these happened to have the properties of a polymerase such that it could either catalyze its own reproduction or, more likely, catalyze the replication of a second large molecule that could contain genetic information in its sequence of monomers. However, for life to begin, the core catalysts and information carriers were necessarily parts of a system that included a container and a transporter and the ability to capture chemical energy from its surroundings. Life today is defined by such systems, one example being the system of enzymatic steps that forms the core metabolic pathway we call glycolysis.

Beyond the simple fact that systems are organized into linear and branched networks, they also are controlled by regulatory processes involving feedback loops. The most common regulation in metabolism is product inhibition, in which the product of an enzyme-catalyzed reaction inhibits the enzyme and in this way avoids having too much of the product accumulate. However, there are also regulatory processes in which a signal is amplified. One of these is the way a single photon of light energy can activate a cascade of chemical reactions that ultimately cause a cell in the retina of the eye to transmit a signal to the brain.

Another property of biological systems is that specific protein components of the system undergo constant interactions. These can now be established by biochemical and genetic methods, resulting in a map of the interactions referred to as an interactome. A challenge for origins of life research is to understand the minimal interactome that will allow life to begin as a functional system of compartments and large molecules capable of catalysis and replication, together with feedback loops that regulate their functions.

10

MULTIPLE STRANDS OF LIFE

In the 1967 movie *The Graduate*, the secret of the good life was revealed to young Dustin Hoffman in a single word: "Plastics!" In fact, life and plastic do share one important property: Both are based on long strings of chemicals that form polymers. In the case of plastic, the chemical units are simple compounds like ethylene (polyethylene) or styrene (polystyrene), but life uses more complex molecules called amino acids to make polymeric proteins, and it uses nucleotides to make nucleic acids like DNA.

At this point in the book, it is time to list the five major gaps in our knowledge that must be filled in before we can understand how life begins:

- Origin of polymers: What were the first polymers of life, and how were they formed?
- Origin of catalysts: How did the first polymeric catalysts appear?
- Origin of replication: What were the first replicating polymers?
- Origin of the genetic code: How did base sequences in nucleic acids begin to code for amino acid sequences in proteins?
- Origin of ribosomes: How did ribosomes emerge as molecular machines that translate nucleic acid sequences into amino acid sequences?

Every one of these fundamental questions concerns polymers, and that's what this chapter is all about.

WHY POLYMERS?

The "mono" in monomer is from the Greek word meaning "single," while the "poly" in polymer means "many." The polymers called proteins typically contain several hundred amino acid monomers, and the polymeric nucleic acids are composed of monomers called nucleotides. The smallest RNA molecules are the transfer RNAs (tRNA) that transport amino acids to ribosomes and contain just 74 to 95 nucleotides. A really large DNA molecule, such as the circular genome of bacteria, contains several million nucleotides.

Why does life depend on polymers, anyway? This might seem to be a tough question, but it's not too hard to understand. By way of analogy, let's think of the letters of the alphabet as unattached monomers. If I tried to answer the above question just by using a mixture of monomers, one possible mix would look something like this:

"rfieisincnpnsnoomoaofoioottieeeeeyeaasaannaaoamsmnlmpmdrrffccsccqldohueu-teysciolthtnsinnw"

I guarantee that the letters above can be arranged into the answer, but they don't make any sense because they are in a random order and could move around if they were not attached to one another. There are 89 places in the string where any of 19 different letters could be placed, so the total number of random sequences I could write is 19^{89}, equal to about 6×10^{13}, which is larger than the number of hydrogen atoms in the observable universe (10^{80} atoms). This is the problem that life would face if it tried to use only monomers without linking them together. There are just too many possible combinations, and none of them are permanent.

Now let's arrange the letters into words, in which certain letters are always linked in specific sequences according to rules:

"specificoftheandpolymersinformationwayinalsofoldasequencemonomerstocontain-causesinthemcan"

A little bit of information begins to creep in, resembling the way that short sequences of monomers can produce specific structures if we impose a set of rules. As your mind tries to puzzle out the words, the lack of spaces between words makes it even more confusing. This means that even the empty spaces used to separate the words are important for conveying information. Later we will see that the information content of DNA also has a kind of spacing that indicates the beginning and ends of genes.

Now I will put the monomeric letters, spaces, and words together in the order that reveals the answer:

"The sequence of monomers in polymers can contain information and also causes them to fold in a specific way."

That's why life depends on polymers. Genetic information is contained in the linear sequences of monomers in nucleic acid polymers, and catalytic functions arise from the sequence of amino acids in protein polymers that cause the protein (or RNA in the case of ribozymes) to fold into virtually unique structures. Once we understand this basic fact, it is difficult to imagine that living systems could exist any other way.

The two kinds of polymers that have won the evolutionary race are proteins and nucleic acids. Proteins are the words of life. When they have a specific spelling (amino acid sequence) and fold into a specific conformation, they can serve either as catalysts or as structural components in a cell. Nucleic acids by the same analogy are the dictionaries of life, containing the coded sequences that tell the cell how to spell the words. By the way, occasionally a misspelling occurs. This can take the form of an added letter, a deleted letter, or a wrong letter. For instance, the word "specific" in the above sentence could be spelled pecific (deleted letter), spechific (added letter), or spacific (wrong letter). It could even be read backwards: cificeps. In each case, the precise meaning is lost. When the letters of DNA are changed it is called a mutation.

Usually the misspelling does not cause any problems because it probably occurs in a non-coding region of the genome, which in the case of humans is most of the DNA (~98%!) that we carry around in our cells. Human beings differ from one another because there are approximately three million points in the genome where the sequence varies, equivalent to one in 1,200 base pairs in our DNA. Each variation is called a single nucleotide polymorphism, abbreviated snp, and pronounced "snip" when geneticists talk to each other in jargon. But occasionally, the mutation occurs in the middle of a gene region, and then problems can arise. A classic example is sickle cell anemia, in which a snp in the hemoglobin gene causes one amino acid in the hemoglobin protein to be replaced by another. The altered form of hemoglobin can still absorb and transport oxygen, but it tends to form a semicrystalline gel when red cells pass through capillaries. The red cells change from their normal biconcave disks to elongated sickle-like shapes, which block the flow of blood in capillaries and produce the symptoms of sickle cell disease.

MONOMERS AND POLYMERS

If we are to understand how life begins, we need to discover simple ways by which polymers can be made from the monomer molecules we think were present in the prebiotic environment. Chapter 4 described the four major components of life, three of which are monomers that can be linked together into polymers. The monomers are amino acids, nucleotides, and simple sugars; and the respective polymers are proteins, nucleic acids, and polysaccharides like cellulose and starch. The fourth component is lipids, the hydrocarbon compounds that can self-assemble into lipid bilayers of membranes. But bilayer structures are stabilized by physical forces, rather than the chemical bonds of polymers, so they fall into a separate class of essential biological structures.

The most important thing to remember about the polymers of life is that virtually all of the chemical bonds linking monomers into polymers are produced by removal of water molecules. Furthermore, the resulting bonds can be broken by the addition of water molecules, a chemical reaction called hydrolysis. All life depends on these two processes. The first reaction synthesizes the essential polymers, and the second reaction breaks down polymers into monomers, which can then be recycled. The cycle of synthesis and hydrolysis is continuous in living cells.

Polymers composed of carbohydrates are also important for contemporary forms of life. They serve as structural components like cellulose or as energy storage compounds like starch. Art Weber, my colleague at NASA Ames in Mountain View, California, has shown that the energy content of a three-carbon sugar called glycolaldehyde can drive a series of reactions that resembles a primitive version of metabolism. Under certain conditions, long polymers are also produced, and Art has proposed that a "sugar world" provided a foundation for metabolic pathways and polymers used by the first forms of life. However, carbohydrate polymers are, at best, very poor catalysts, and they cannot store genetic information. For this reason, most research on the origins of life has focused on polymers of amino acids and nucleotides, which can be linked together into proteins and nucleic acids.

POLYMERS COMPOSED OF AMINO ACIDS

While I was a chemistry major at Duke University, I had my first introduction to the idea that life is composed of polymers. I was working as an undergraduate researcher in a botany laboratory supervised by Aubrey Naylor. Bob Barnes, a tall, skinny graduate student also working in the lab, was analyzing certain plant proteins to establish their amino acid content. I expressed curiosity about what he was doing, and Bob agreed to let me try the technique. "What protein do you want to use?" he asked. I had heard by then that most biological substances contained proteins, so I volunteered a fingernail clipping. "Perfect." Bob dropped the clipping into a test tube of concentrated hydrochloric acid and put it on a heating pad. "This will take all night, so we'll run the chromatogram tomorrow."

The next morning when I arrived at the lab, I checked the test tube. The fingernail had completely disappeared, dissolved by the combination of strong acid and heat that hydrolyzed the peptide bonds. I watched as Bob carefully placed a drop of the clear solution at the corner of a square sheet of paper about 18 inches across. He then draped the sheet of paper from a tray in a large tank so that a mixture of butanol and acetic acid could slowly creep down, wetting the sheet from top to bottom. A few hours later, after the sheet was almost completely wet, he took it out of the tank, dried it, and turned it sideways, draping it again from a tray but this time in a mixture of smelly phenol and water.

Toward the end of the day, the sheet was again wet from top to bottom. Bob removed the wet sheet, dried it in a hood, and sprayed it with a solution of a dye called ninhydrin, which reacts with amino acids to produce purple, blue, and pink colored compounds. After briefly heating the sheet in an oven, he handed it to me. I was amazed! The sheet had become a work of abstract art, filled with multiple colored spots, each representing an amino acid used by the cells in my finger that linked amino acids together to form keratin, the tough protein of hair, skin, and, of course, fingernails.

The technique Bob used was called two-dimensional paper chromatography, state of the art at the time, but no one would use it today. Instead, we would inject a tiny volume of the mixture into an instrument in which high performance liquid chromatography (HPLC for short) or capillary electrophoresis separates the amino acids and produces an analysis in a few minutes rather than a day. We could use virtually any protein, such as the albumin of egg white or the hemoglobin of blood, and get a selection of the 20 amino acids used by all life today. What gives each protein its distinctive properties is not just the fact that they are composed of amino acids, but rather how much of each amino acid is in the protein and the sequence of amino acids in the linear polymer.

Amino acids are present in carbonaceous meteorites, and they are synthesized in Miller-Urey simulations of prebiotic Earth, so it is reasonable to conclude that they were among the ingredients composing the dilute organic solution in ponds, lakes, and seas four billion years ago. The next question is: How did they link up into the first polymeric compounds we now call proteins? To address this question, we need to know a little more biochemistry related to amino acid monomers and polymers.

Let's begin with amino acid structure, which is as simple as taking a carbon atom and attaching the amine ($-NH_2$) and carboxyl group ($-COOH$) that define all amino acids. Then we also add a third chemical group (abbreviated $-R$) that defines each specific amino acid and finally, a hydrogen atom ($-H$) to fill in the fourth bond that is available on a carbon atom. The nature of the $-R$ groups give amino acids specific chemical properties when they are part of a protein chain. For instance, the $-R$ group could be $-CH_2-COO^-$ for aspartic acid or $-CH_2-CH_2-COO^-$ for glutamic acid, which adds a negative charge. Or it could be $-CH_2CH_2CH_2CH_2-NH_3^+$ in lysine, adding a positive charge. The $-R$ groups of other amino acids such as alanine, valine, and leucine are hydrocarbons, making them hydrophobic, and if the $-R$ group is a hydroxyl ($-OH$) as in serine and threonine, they are hydrophilic. A special case is the sulfhydryl group of cysteine ($-SH$) which can link to another cysteine by forming a disulfide bond ($-S-S-$). The disulfide linkage is the only way that two peptide chains can be chemically bonded together in a protein. For instance, insulin is composed of two short peptides called the A and B chains, and these are held together by disulfide bonds in the active hormone.

The unique feature of amino acids is that the amine and carboxyl groups can lose a water molecule to form linkages called peptide bonds (Figure 19). These bonds are produced in cells by a very complex process involving ATP as an energy source and multiple enzyme-catalyzed reactions that link an amino acid to a transfer RNA.

HOW ARE PROTEINS SYNTHESIZED BY LIVING CELLS?

Earlier in this chapter, I used letters and words to illustrate how information can be encoded in a sequence of monomers in a polymer. In order to introduce the mechanism by which a ribosome reads the sequence of bases in messenger RNA, I will use another

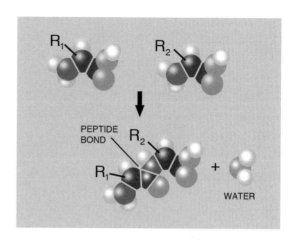

Synthesis of a peptide bond between two amino acids is equivalent to loss of a water molecule. The actual removal of water occurs when ATP is synthesized in mitochondria from ADP and phosphate. The ATP then activates the tRNA-amino acid complex that is used by the ribosome to produce a peptide bond.

analogy, this time to musical notation. When I taught introductory biology courses in the early 1980s, the classes were large—several hundred students. What professors discover early in their career is that lecturing to large classes is a performance, unlike the conversation that we can have in smaller classes. So I tried to capture the attention of students by laying out knowledge of biology in memorable ways, rather than just strings of words in a dry lecturing style. When we got to translation, I realized that I could portray gene sequences and the function of ribosomes by making an analogy to music. In other words, the sequence of notes on a sheet of music (like the sequences of letters in words or words in a sentence) carries information. The first sequences of bases in genes encoding for proteins were being published in the early 1980s, so I decided to transform the base sequence of human insulin into music and play it for my students.

The rules for doing this are simple. It happens that three of the four bases have the same abbreviation as musical notes: A, C and G (adenine, cytosine, and guanine). If I used the musical note E as the code for thymine (T), the result was a four-note scale that made musical sense and sounded good to the ear. On a piano, if you play the notes ACEG together, it produces a nice A minor seventh chord, and CEGA is a jazzy C major sixth chord.

I could now play the entire sequence of bases in the insulin gene as a melody. For instance, TTT GTG AAC CAA are the triplet codes for phenylalanine, valine, asparagine, and glutamine—the first four amino acids—and these translated into the musical notes EEE GEG AAC CAA. If you would like to hear some of the melodies transcribed from DNA sequences, the music is available on the Internet. (See "Sources & Notes" for this chapter.) There are two main points to be made here. First, understanding that a linear code of notes in sheet music can be translated into actual music by a musician gives a good sense of how a linear code in a nucleic acid like DNA can be translated into a protein by a ribosome. By analogy, the sheet music is the gene and the musician is the ribosome that translates the sequence of notes into the sound the audience hears. Furthermore, the tempo that a musician plays the insulin gene is similar to the rate of amino acid addition to a growing protein chain, which is in the range of a few per second.

Now let's talk about transcription, translation, and ribosomes. This is not a biochemistry textbook, so I will only sketch the main features of protein synthesis in living cells today. In 1957, Francis Crick, the co-discoverer of the DNA double helix structure, proposed what he called the Central Dogma. Enough information had accumulated by that time so that it could be integrated into a coherent picture of protein synthesis that is now in every college textbook of biochemistry and even presented to high school students. What we know now is that genes direct the synthesis of proteins, and genetic information is stored in the sequence of nucleotides in DNA. To express a gene, the double helix of DNA is unwound and an enzyme called RNA polymerase transcribes a specific portion of one of the two strands into a single strand of RNA, using the DNA strand as a template. In this way, the order of nucleotides in the DNA is precisely transcribed into the sequence of nucleotides in the RNA, which is referred to as messenger RNA (mRNA). The mRNA has what are called codons, each composed of three bases, and each codon codes for a specific amino acid. This is referred to as the genetic code, which is virtually universal in all life today. In other words, three uracil bases in a row (UUU) comprise the codon for phenylalanine, three adenines in a row (AAA) are the codon for lysine, and so on for all 20 amino acids.

Ribosomes synthesize proteins from amino acids, and the sequence of amino acids in the protein is dictated by the sequence of codons in the mRNA. Ribosomes are complex molecular machines composed of 56 total proteins, 22 in a small subunit and 34 in a large subunit. In bacteria, the small subunit has a single strand of ribosomal RNA called 16S rRNA, and the large subunit has a strand of 5S and a strand of 23S rRNA. The S units (as in 23S) can be a little confusing at first, but they simply have to do with how fast a molecule moves downward through a fluid exposed to a strong centrifugal force. The S units are abbreviations named to honor Theodor Svedberg, who was awarded the Nobel Prize in 1926 for his invention of the ultracentrifuge. This device spins a rotor so fast that a tube in the rotor experiences a force 100,000 times that of gravity, so that even large molecules in solution move toward the bottom of the tube. The main thing to remember is that larger molecules like 23S rRNA move downward faster than smaller molecules such as 5S rRNA.

Over the past 30 years, several research groups around the world began heroic attempts to determine the exact structure of ribosomes, heroic because this would be the largest and most complex biological structure ever to be established at the molecular level of resolution. The only way to do this was to crystallize ribosomes (no small feat in itself) and then use X-ray diffraction to analyze the crystals. Four research groups began to make significant progress over the years, and little by little the structure of the ribosome began to appear in publications from four leading laboratories. While I was writing this book, three Nobel Prizes were awarded to the investigators who were bold enough to enter the race. These included Ada Yonath at the Weizmann Institute of Science in Rehovot, Israel, Thomas Steitz at Yale University, and Venkatraman Ramakrishnan, who works at the MRC Laboratory of Molecular Biology in Cambridge, England. If the prize could have been split four ways, most would agree that Harry Noller at the University of California, Santa Cruz, would also have been honored for his studies of ribosome structure.

Figure 20 shows a ribosome in action. The mRNA travels to one of the ribosomal subunits, which binds the strand, and then the second subunit is attached. Meanwhile,

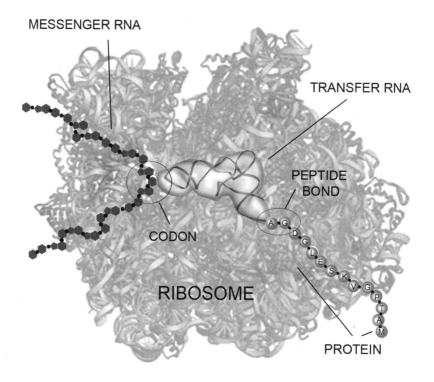

FIGURE 20

Ribosomes are composed of a large and a small subunit, shown here in the background as a light gray tangle of protein and nucleic acid chains. To begin the protein synthesis process, the two subunits bind to a strand of messenger RNA to form an active ribosome. A single-stranded messenger RNA molecule is shown in the drawing, moving through the site of active protein synthesis on a ribosome, where a transfer RNA molecule is in the process of delivering an amino acid to a growing polypeptide (protein) chain. As each triplet moves into position, a transfer RNA with an amino acid clicks into place, delivers its amino acid, then leaves to make room for the next tRNA. The amino acid is covalently linked through a peptide bond to the growing polymer that will become a protein, after which the mRNA clicks to the next codon. (See references in Sources and Notes for Chapter 10 to view animated simulations of this amazing process.)

another enzyme uses the energy in ATP to covalently link amino acids to a third kind of RNA called transfer RNA (tRNA). There is one enzyme and one kind of tRNA for each kind of amino acid, and the other end of the tRNA has an exposed sequence of three bases. Protein synthesis is initiated when the three-base sequence on a methionine tRNA binds to three bases of the mRNA on the ribosome, and the amino acid is transferred to the ribosome itself. Almost like a machine, the mRNA moves along the ribosome to expose the next codon. Then there is a pause of less than a second while other tRNAs diffuse to the newly exposed codon, until finally one fits according to the universal genetic code. The amino acid attaches through a peptide bond to the methionine, and the process repeats until a complete protein molecule is synthesized, typically several hundred amino acids in length. Because ribosome function is so central to life, a number of animations are now available that help visualize what actually happens during translation. (See "Sources and Notes" for Chapter 10.)

Just imagine how mysterious this process was 60 years ago. Working out the mechanism was a combined effort of scientific genius, and Nobel Prizes were sprinkled liberally among the scientists who made the major advances. The question now is: How did it all begin? More specifically, what were primitive mechanisms of replication, transcription, and translation? And how did the genetic code arise with a triplet code for each of the 20 amino acids? We don't know yet, but step by step over the next 50 years, I am optimistic that answers will be forthcoming, just as they slowly became apparent after the double helix of DNA was solved and led to the Central Dogma of biology.

HOW WERE POLYMERS SYNTHESIZED ON PREBIOTIC EARTH?

"May you live in interesting times," says the ancient Chinese curse, and times certainly must have been interesting for Aleksandr Ivanovich Oparin, who was 23 years old when he graduated from Moscow State University in 1917. Lenin and the Bolsheviks had just seized power, the Czar and his family were imprisoned and then assassinated a year later, and the war between Red and White Russia began. But the young scientist managed to avoid getting caught up in the chaos, perhaps because he was thinking about something far more interesting than politics: How did life begin? Oparin began a serious effort to put his thoughts together in 1922, and his book *The Origin of Life* was published in 1924 as a Russian-language monograph.

After Louis Pasteur had concluded in 1864 that life always came from pre-existing life, no one had given serious thought to the problem of the origins of life, which is why Oparin's book stands out so clearly as a milestone in the scientific landscape. Oparin's basic idea was that there must be a way for living systems to arise from nonliving chemicals on early Earth. In 1928, J. B. S. Haldane independently published a short, prescient essay expressing similar ideas, and now we refer to their founding concepts as the Oparin-Haldane scenario. What was known at the time is that the unit of life is the cell, and proteins are structural components. Cells did not appear to be microscopic bags of watery stuff, but contained protoplasm similar to jelly. It was also known that colloids—sub-microscopic aggregates of compounds like starch and gelatin—tended to be water soluble when dilute, but they formed gels when concentrated. The gels sometimes produced microscopic spherules called coacervates, and Oparin adopted the term to describe what he thought the first forms of life might have been like.

Thirty years passed, and Oparin became one of the premier scientists of the Soviet Union. In 1957, under the auspices of the Academy of Sciences of the USSR, Oparin helped organize a conference in Moscow that attracted some of the world's leading scientists, including Stanley Miller and his mentor Harold Urey, who had won the 1935 Nobel Prize for the discovery and isolation of deuterium. Linus Pauling and Wendell Stanley, both Nobel Prize winners, were also there, as well as future Nobelists Melvin Calvin, Ilya Prigogine, and Peter Mitchell. The proceedings of the meeting were published by Plenum Press as a 700-page book, and it is fascinating to read papers written 50 years ago, watching these brilliant intellects wrestle with the question of life's origin.

Sidney Fox, at the time the new director of the Oceanographic Institute at Florida State University, was also a participant. Fox thought he knew how life began and was one of the first scientists to test the idea that polymers resembling proteins could have appeared spontaneously on prebiotic Earth. In 1958, Fox and his colleague Kaoru Harada reported that heating dry amino acid mixtures to 180°C produced polymers resembling proteins. The dry heat essentially baked out water molecules so that chemical bonds linked the amino acids together into polymers they called thermal condensation products, or proteinoids. A few years later, Fox and his collaborators found that under certain conditions, the proteinoids assembled into microscopic spherical balls, which they referred to as proteinoid microspheres. These were similar to the coacervate structures described by Oparin, and the fact that they were composed of amino acid polymers made them a more attractive model of a prebiotic protocell than the mixture of gelatin and gum arabic that Oparin had studied. Furthermore, dry heat seems like a very plausible energy source for driving condensation reactions. One imagines, as Fox did, that extensive volcanic activity on early Earth would provide many opportunities to produce interesting polymers.

For a while, there was a consensus that Fox was on the right track, and images of proteinoid microspheres appeared in biology and biochemistry textbooks in the 1970s. But in the 1980s, proteinoids were abandoned in favor of RNA. What happened? Why hasn't dry heat been accepted as the most likely energy source to drive polymerization leading to the origin of life? There are several reasons. First, heat by itself is not a true energy source. Nothing happens in a hot solution of amino acids except that the acids slowly break down into simpler compounds. A better way to think about the role of heat in a reaction is that it adds activation energy to potential reactants. If monomers can undergo a spontaneous polymerization reaction, higher temperatures will cause the reaction to go faster. When the conditions employed by Fox and his coworkers caused amino acids to polymerize into proteinoid polymers, the actual chemical energy supplied to the reaction was the fact that the reactants were dry so that a variety of chemical bonds could form between the reactants when water molecules were driven off the monomers by the elevated temperature. As a result, covalent bonds formed between the monomers to produce long polymeric chains.

Although this is a plausible way to drive prebiotic condensation reactions, there are significant caveats regarding the kinds of chemical bonds that form when dry material is heated. A mixture of hot chemicals can undergo multiple different reactions, all producing chemical bonds. If the temperature is high enough, the result is a brown or black substance commonly referred to as "tar." This is one of the chief criticisms of Fox's approach and more generally, of the "shake-and-bake" approach to modeling the origin of life: The reactions and products are nonspecific. The particular mixture of amino acids Fox used does form a few peptide bonds, but other chemical bonds are also produced that never contribute to the molecular structure of biological proteins. Another problem is that no one has yet found a plausible way for proteinoids to carry and transfer genetic information. Nucleic acids can do this, so we now think that the first forms of life might have used RNA as the first functional polymer rather than proteins.

In his enthusiasm for thermal proteinoid microspheres, Fox became convinced that he had discovered how life began, and he made claims that went far beyond what most scientists were willing to accept as plausible. Toward the end of his career, Fox was proposing that proteinoid microspheres were alive in every sense of the word—growing, reproducing by budding and fission, and even exhibiting a primitive form of nervous activity. This is a fascinating scientific saga, with many twists and turns, illustrating how scientists can be led astray when they try to read too much into their observations. For the most part, the shake-and-bake approach to prebiotic polymer synthesis has been abandoned in favor of defined chemical systems that work in aqueous solutions. For instance, Luke Leman, Leslie Orgel, and Reza Ghadiri at the Salk Institute and the Scripps Research Institute in La Jolla, California, recently investigated carbonyl sulfide (COS) as an activating agent. COS is known to be present in volcanic gases, and has the same chemical structure as COO (carbon dioxide) except that a sulfur atom replaces one of the oxygen atoms bonded to carbon. In 2004, these authors reported that COS can chemically activate amino acids so that they react to form peptide bonds. In 2006, the same group discovered that carbonyl sulfide also activates phosphate to form high-energy anhydride bonds if amino acids are present. These reactions lead to fairly specific products and take place in aqueous solutions, which chemists much prefer to the uncontrolled reactions occurring in dry hot mixtures of amino acids.

Even though the COS activation is an elegant example of how volcanoes can be linked to prebiotic chemistry, true polymers have not been synthesized. So far only peptide bonds that link a couple of amino acids into dimers have been synthesized. I think there may still be discoveries to be made in the shake-and-bake approach, and in Chapter 14, I will describe how a lipid matrix can organize nucleic acid monomers in such a way that long polymers are produced by cycles of dry heat and wetting.

PROTEIN FOLDING IS ESSENTIAL

It is not enough just to make polymers of amino acids. As noted earlier, the sequence of amino acids in a peptide chain, or of nucleotides in a nucleic acid strand, determines the way that the polymer folds. In proteins, there are three levels of structure related to folding. The primary structure is simply the sequence of monomers in the molecule, the secondary structure is the way that a protein strand forms helical and sheet-like patterns, and the tertiary structure is the way that helices and sheets fold up into stable three-dimensional conformations that represent the functional structure of the protein.

How do we know all this? After all, we are talking about the specific arrangement of thousands of atoms that compose a single protein molecule. At this point I want to describe a specialized technique called X-ray diffraction, which is the single most important method for establishing molecular structures at near-atomic resolution. The method relies on the fact that waves interact to produce constructive and destructive interference. This effect can be seen in a backyard swimming pool when it is not in use, so the water surface is smooth. If something about the size of a marble is dropped in the pool, a pattern of concentric circular waves is produced. But if we drop in two marbles at the same time, about two feet apart, the two sets of circular waves pass through each other. If you watch carefully you will see a series of higher spots where the waves come together and reinforce each other (in phase, or constructive interference) and other areas where the crest of one wave meets the trough of another, and they cancel each other (out of phase, destructive interference).

Those are waves in a swimming pool, but you can also see interference patterns with waves of light. All it takes is a laser pointer and a fine-grained material that has a repeating structure in it, such as a CD or DVD disk. Shine the laser on the disk so that the light reflects onto a white wall and you will see not just a single spot that a mirror would give, but multiple spots, some of which have a second smaller pattern of spots around them. These are diffraction patterns that are produced when the laser light reflects off the data track on the recording surface of the disk, which is a single spiral over 3 miles long (!) with a repeat distance of about 1.6 microns between the lines of data that compose the spiral. When the photons from the laser reflect off the lines, they produce a series of light rays, each taking a slightly different distance to get to the wall. All of the rays are exactly the same wavelength, so the distance that some of the rays travel is such that they are in phase and reinforce each other, which produces brighter spots called constructive interference. Other rays are out of phase, which leads to destructive interference and produces the dark regions between the spots. From the known wavelength of light and the angle of a given spot with respect to the central beam, a simple formula gives the distance that separates each of the repeating units in the disk.

Now imagine that we use not visible light, but instead the light we call X-rays, which have wavelengths 1,000 times shorter than visible light. The X-rays in a modern instrument are produced when a stream of electrons hits the surface of a rotating tungsten-rhenium target. The wavelength of the X-rays is in the range of the distances between atoms in the crystal, which means that X-ray crystallography can achieve atomic resolution. As a test, we might first send the X-ray beam from the instrument through a piece of plastic such as polyurethane, which is composed of a tangled mass of polymeric molecules in no particular order. In older instruments, the beam would fall on a photographic plate for a few minutes, but today a CCD (charge-coupled diode) is used, similar to those in digital cameras. The beam produces a bright point surrounded by a diffuse haze. The X-rays have been scattered by the polymer molecules, but there is no orderly arrangement, so the scattering just produces a circular fog around the beam, something like looking at a flashlight bulb through a piece of frosted glass.

But a protein crystal has a high degree of order, so the X-rays of a beam passing through are reflected off the electron shells that compose different planes of atoms within the crystal. The rays undergo the same in-phase and out-of-phase interference that the laser light does. The crystal is slowly rotated while exposing a photographic plate to the beam (or in modern instruments, a CCD array, similar to that in a digital camera), and the reflected beams produce a very detailed interference pattern of hundreds, even thousands, of spots on the plate. From the spacing and density of the spots, the angle each spot makes with respect to the central beam, and the degree of rotation of the crystal, the precise, three-dimensional crystal structure can be calculated with atomic resolution. In the early days of protein crystallography, the calculations were done by hand and required months, even years, of work. The first proteins to be resolved were myoglobin and hemoglobin, and won a Nobel Prize in 1962 for Max Perutz and John Kendrew, who did the work at Cambridge University. Today the calculation is carried out by computers, and virtually anyone can solve a crystal structure in a few days as long as they can get the protein to crystallize.

In the 1930s, William Astbury used X-ray diffraction to study changes that occurred in wool when it was stretched, and guessed at helical and sheet structures. However, the precise alpha helix concept was the original idea of Linus Pauling, who understood that hydrogen bonding between carbonyl oxygens and amide nitrogens in the peptide chain could stabilize such universal structures. Pauling tells the story that he was at home in Oxford, England, during a visit in 1948 and had come down with a cold. To ward off the boredom of lying around in bed, he began to think about the structure of keratin, the protein of hair that has a relatively simple amino acid composition with lots of glycine and alanine. Pauling wrote down the known linear structure of a peptide chain and tried folding the paper in various ways. He suddenly realized that if the chain was wound up into a spiral structure that repeated every 3.6 amino acids, a beautiful helix was produced that maximized hydrogen bonding possibilities and minimized internal strains. Pauling and Corey went on to publish exact structures of the alpha helix and beta sheet in 1951, matching them to X-ray diffraction data and initiating a revolution in our understanding of protein structures.

Despite the beauty of secondary structures, they are not the end of the story. The catalytic function of an enzyme molecule can only begin after it folds into what is called a tertiary structure. Figure 21 shows the folded structure of an enzyme called ribonuclease, which catalyzes the hydrolysis of RNA into monomer nucleotides. Two of the secondary structures are clearly evident. An alpha helix is shown as a spiral, and a beta sheet is shown as neighboring ribbons. Ribonuclease binds to ribonucleic acid and catalyzes hydrolysis, a kind of digestion process in which water molecules are added to phosphodiester bonds so that the nucleotide monomers are released. All forms of life use ribonuclease for this purpose, and bacterial ribonuclease is everywhere in the environment—so much so that we must wear latex gloves in the laboratory if we are going to work with RNA. Otherwise a few of the enzymes on our fingers will inevitably get into the samples and we end up not with pure RNA, but instead with little pieces of partially digested material.

The amazing thing about protein folding is that the specific sequence of amino acids in the chain causes that chain to fold into the same tertiary structure virtually every time. For instance, Figure 21 also shows what happens when ribonuclease is heated in water. The heat energy added to the molecule causes it to unfold, or become denatured, but the ribonuclease is able to refold into precisely the same original structure upon cooling. The most important aspect of folding is that when the correct folds spontaneously occur, groups of amino acids are brought together to produce what is called an active site, the catalytic center of all enzymes. If a substrate molecule happens to enter the site, the amino acids of the active site interact with the molecule and alter its electronic structure

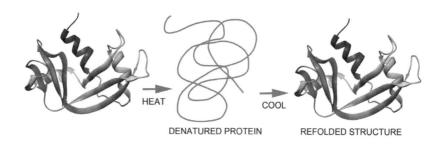

HEAT   COOL

DENATURED PROTEIN   REFOLDED STRUCTURE

FIGURE 21

Ribonuclease catalyzes the hydrolysis of RNA molecules into its component nucleotides. This enzyme is a relatively simple protein molecule that has three alpha helical structures and a beta sheet. Ribonuclease can be denatured by heating, but then spontaneously folds back into its active tertiary structure when cooled.

in such a way that the molecule more readily becomes a reactant in a specific chemical reaction. This can be as simple as the breakdown of H_2CO_3 into H_2O and CO_2 (a reaction catalyzed by carbonic anhydrase) or as complex as the synthesis of the phosphodiester bonds of nucleic acids by a polymerase.

What does all this have to do with the origins of life? In a word: everything. Catalytic activity is essential for life to begin, and only a folded structure can be a catalyst—either a small protein or, as discussed next, a strand of RNA in the form of a ribozyme. This brings us to three of the deepest questions related to life's beginning:

- How were amino acids linked together into peptide chains from a mixture of amino acids?
- How were the first correctly folded protein catalysts selected from among the vast number of possible random sequences?
- How was homochirality imposed on the first protein-like polymers? Correct folding can only occur if the amino acids are entirely homochiral, as are the L-amino acids in all existing organisms.

We don't have answers to these daunting questions yet, but we can of course set up working hypotheses as possible explanations and then test them experimentally.

NUCLEIC ACIDS

The other primary polymers of life are the nucleic acids, represented by DNA and RNA. These are so central to our story that I will take some time to describe their history and structure. DNA was discovered by Johannes Friedrich Miescher in 1871, long before its significance as the carrier of genetic information was established. Miescher was Swiss, but he traveled to Tubingen, Germany, to work with his mentor, Felix Hoppe-Seyler. At the time, scientists who would now be called biochemists were attempting to isolate and investigate the chemical components of living organisms. Proteins had already been discovered and named from the Greek word *proteios*, which has the sense of being first. (Similar "first" words include "proton," "Amoeba proteus," and "prototype.") Miescher decided to study the chemical composition of nuclei, the small round structures in most cells that were revealed by newly developed techniques of staining and microscopy. The microscope was being used to study various tissues, and Miescher knew that white cells called leukocytes were present in the pus that exuded from all-too-common infected wounds at hospitals. (We now know that leukocytes accumulate at sites of injury to help combat bacterial infections.) To prepare leukocytes, he rinsed them off bandages that were loaded with pus and then broke up the cells to release the nuclei, which settled to the bottom of the beaker. These he treated with an alkaline solution that precipitated proteins, which were discarded, followed by an acidic solution that precipitated a slimy clear material that he named nuclein. No one had any idea what the stuff was, but it did earn a professorship for Miescher at Basel, Switzerland, where he spent the rest of his career.

The next person to work on nuclein was Albrecht Kossel at the University of Heidelberg, who was able to determine that it was not a protein, as previously thought by Miescher, but instead was a mixture of a protein he named histone and another substance composed of a mix of purines (adenine and guanine) and pyrimidines (thymine, cytosine, and uracil). Kossel also discovered that carbohydrates were present, along with phosphate. Considering the limited technical approaches available at the time, this was remarkable work, and Kossel was rewarded with a Nobel Prize in 1910. The next major advance was made by Phoebus Levene, who worked in New York City at the Rockefeller Institute of Medical Research (later to become Rockefeller University) in the early 1900s. By this time, it was understood that the phosphate in nuclein had acidic properties, so it was now called nucleic acid. Levene discovered that purified nucleic acid was a polymer composed of repeating units he called nucleotides, which were linked together through ester bonds between the deoxyribose sugar and the phosphate. With that discovery, the stage was set for perhaps the single most important advance in our understanding of life on Earth, that the sequence of bases in DNA could encode information required to reproduce living organisms, from bacteria to human beings.

The rest of this story does not need retelling, and the functional role of DNA as a genetic material can be found in any college-level biology textbook or treatises on scientific history. But we do need to complete the story related to DNA's structure. Six people loom large in this history: William Astbury, Maurice Wilkins, Rosalind Franklin, Linus Pauling, James Watson, and Francis Crick. Astbury was at the University of Leeds, and is not as well known as the other five, but he deserves credit for his pioneering work in which he applied X-ray diffraction to fibrous biological material, such as the keratin protein of wool, in which he detected a repeating pattern that inspired Pauling's alpha helix. In 1937, Astbury was given a purified sample of dry DNA prepared from the thymus gland of a calf. He sent a beam of X-rays through the material and was delighted to see a pattern appear on the photographic plate. The pattern indicated that there was a certain amount of order in the DNA, and Astbury published a paper in which he noted that the purine and pyrmidine bases in DNA were present as stacks with a repeating distance of 3.4 angstroms. Then World War II interrupted progress, and not much more was done until Rosalind Franklin began her work in John Randall's laboratory in 1951 at King's College London. She did obtain clear patterns that had clues, but she did not want to guess at DNA structure. Franklin was confident that by doing X-ray diffraction carefully, it would be possible to deduce the exact structure.

Two other young scientists, however, loved to make guesses. Instead of experimental observation, they thought it should be possible to propose a structure using molecular models. Their names have now become household words, at least in households of molecular biologists. James Watson was a post-doctoral student who had been attracted to work in Cambridge, and Francis Crick was attempting to complete his PhD. You can read this story in Watson's autobiographical book *The Double Helix,* where he tells of getting close to a structure in their thinking and then getting a glimpse of Franklin's diffraction patterns, which confirmed that they were on the right track. Figure 22 shows the remarkable structure they came up with, in which the bases form hydrogen-bonded pairs in the center of a double helix. Adenine binds to thymine, and guanine binds to cytosine, now referred to as Watson-Crick base pairs. Phosphate and deoxyribose are linked through phosphodiester bonds into two strands that form the stable polymeric structure of the double helix illustrated in Figure 22. Watson and Crick published their idea in *Nature* in 1953, and having a correct structure for DNA marked the beginning of a revolution in molecular biology that continues today. For their contributions, Watson, Crick, and Wilkins were awarded the Nobel Prize in 1962. Sadly, Rosalind Franklin was a victim of cancer, and the significance of her contribution was only recognized after her death.

By the way, calf thymus remains a major source of DNA for researchers. If you want some DNA to play with, send a check for $61.20 to the Sigma-Aldrich Corporation and

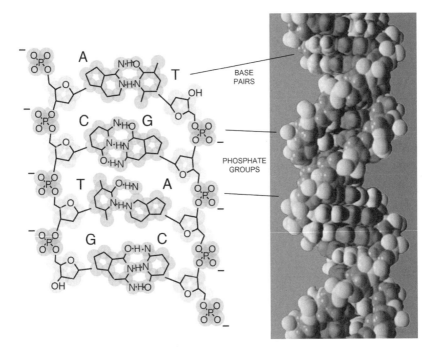

FIGURE 22

The DNA double helix. The stick model on the left shows the two backbones composed of phosphate and deoxyribose, with the Watson-Crick base pairing in the center, in which adenine forms hydrogen bonds with thymine, and guanine with cytosine. The three-dimensional model on the right illustrates how the two strands form a helical structure called the double helix. Image courtesy of the Center for Information Technology, National Institutes of Health.

you will get a small glass bottle containing 1 unit (a few milligrams) of a purified white stringy substance. In principle, you could use this DNA to reproduce your own calf. In fact, George Church at Harvard recently suggested that it might someday be possible to reproduce a Neandertal when the complete genome becomes available, but he cautions that you will also need $30 million to purchase a modern laboratory facility and hire a dozen molecular biologists who are willing to give it a try. Another Nobel Prize awaits the person who reproduces a species using only its DNA, but for ethical reasons, I doubt that it will ever be a member of an extinct human species. Most likely it will be a laboratory rat or perhaps a mammoth.

So, where are we now? First we should understand the basic structure of a single strand of DNA and RNA. This is not too difficult. The backbone is simply alternating phosphate and sugar groups that we can abbreviate $-P-S-P-S-P-S-P-S-P-S-$ and each sugar is either a deoxyribose in DNA or a ribose in RNA (Figure 22). Bases called purines and pyrimidines are attached to the sugars through an unusual $-C-N-$ bond. They are called bases because the pure compounds dissolved in water produce an alkaline (basic instead of acidic) solution. The purines, adenine and guanine, and the pyrimidine cytosine are present in both DNA and RNA, but thymine is in DNA, and uracil is in RNA.

CAN NUCLEOTIDES FORM NUCLEIC ACIDS IN THE ABSENCE OF ENZYMES?

Nucleic acids are at the center of life processes, and cells today use complex enzyme-catalyzed processes to produce strings of RNA and DNA from activated monomers like ATP, TTP, UTP, GTP, and CTP. But some of my colleagues working on the origins of life think that an early form of life began to use RNA long before enzymes could have been present. The question is: How could such a complicated molecule have been produced spontaneously in the prebiotic environment?

The way that the scientific method approaches difficult problems like this is to put together a set of questions that represents a kind of cage. The cage surrounds the unknown answer, and each question is a bar of the cage that helps to constrain our thinking about possible answers. Let's first consider just the nucleotides. To make the DNA and RNA used by life today, we would need a source of five nucleobases, two sugars, and phosphate—all mixed together in one place and sufficiently concentrated to undergo synthetic reactions. The nucleobases include two different purines (adenine and guanine) and three pyrimidines (cytosine, thymine, and uracil), representing the five different bases of DNA and RNA. So the first question is: Where did these nucleobases come from? One of the bases—adenine—has been found in meteorites in small quantities (parts per million by weight), and we know that the other bases can be synthesized from hydrogen cyanide in laboratory simulations of prebiotic chemistry, so perhaps small amounts of the bases were present in the dilute solution of organic compounds.

But how did they get together in one place and become sufficiently concentrated to enter into reactions that produced nucleotides?

The second question concerns a source of two different sugars—deoxyribose and ribose. Where did they come from? Again, simulations show that small amounts can be synthesized from formaldehyde, along with many other sugars, but how did they get together with nucleobases in one place? There is also the question of chiral purity. The two sugars had to be mostly D- or L-forms to produce a nucleic acid polymer (the D-form is in all life today), so how did one form get incorporated into the nucleotide monomers?

Another question involves a source of phosphate. We know that phosphate is relatively uncommon, and at alkaline pH ranges, it tends to be react with calcium to produce the insoluble mineral called apatite. How did enough soluble phosphate get together with the bases and sugars to produce nucleotides?

The last question concerns the actual reactions by which nucleotides are synthesized. The most challenging bond to form is the bond between a specific nitrogen in the bases and a specific carbon in the sugar, which produces a nucleoside. This is such difficult chemistry that it has been referred to as the "nucleoside problem."

Even assuming that all of these components somehow were mixed together under conditions that promoted the synthesis of nucleotides, we face the problem of polymerizing them into nucleic acid molecules, presumably RNA, that were sufficiently long so that they could fold into catalytic conformations of ribozymes. Last, but not least, RNA is relatively unstable and is constantly breaking up into smaller fragments by hydrolysis in a matter of hours to days, so any synthetic process will need to occur at a rate fast enough to keep up with hydrolysis.

If this sounds like a tough nut to crack, well, it really is. The set of questions above brings us right to the edge of our knowledge about how life begins. But there is one possible way through the thicket. My knowledgeable colleagues who have added up the questions, as I have above, agree that it is highly improbable that the proteins and nucleic acids used by life today could have formed spontaneously under prebiotic conditions. What is the alternative? Something simpler, of course. Instead of all 20 amino acids, perhaps the earliest forms of life could get along with fewer, such as the six amino acids that are the most common in Miller-Urey products and carbonaceous meteorites. This means that life would also require fewer nucleobases for a genetic code. For instance, the purine adenine is a common product of HCN polymerization, and uracil is the simplest pyrimidine and is relatively stable. These two bases could code for four amino acids as a doublet code (AA, UU, AU, UA) or for eight different amino acids as a triplet code (AAA, UUU, AAU, UUA, AUA, UAU, AUU, UAA). The fact that A forms base pairs with U also allows a double helix to be produced, which would be necessary for replication. Instead of a complicated sugar such as ribose, maybe the earliest life forms incorporated a simpler linking molecule like ethylene glycol ($HO-CH_2-CH_2-OH$) or glycerol ($HO-CH_2-CHOH-CH_2-OH$). If so, this would also solve the chirality question because these molecules have no chiral centers.

HOW CAN NUCLEIC ACID POLYMERS BE SYNTHESIZED IN THE PREBIOTIC ENVIRONMENT?

Like politics, science can be described as the art of the possible. Even though we don't know yet what the primordial version of a nucleic acid was, we can at least work with molecules that we have in hand, the nucleotides, and see what we can learn in general about polymerization reactions. In principle, it should be easy to make nucleic acids simply by polymerizing nucleotides. All we need to do is to get the molecules close enough together so that the phosphate groups nestle up to the sugar groups in such a way that one water molecule falls off and an ester bond is produced as shown in Figure 23. The result is the characteristic linking bond of all nucleic acids, called the phosphodiester bond. It sounds easy, but there are several reasons why it doesn't work. First is that the formation of an ester bond is energetically uphill, so it does not happen spontaneously. Second is that the nucleotides are odd-shaped molecules, and there is no guarantee that they will fall together in precisely the right way so that the phosphate and sugar groups come together.

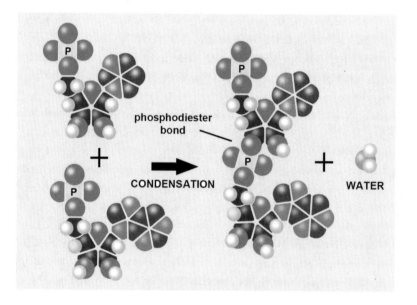

FIGURE 23

A phosphodiester bond between two nucleotides is produced by a condensation reaction with loss of a water molecule. The bond is produced when the –OH group on the number 3 carbon of a ribose sugar attaches to a phosphate group esterified to the number 5 carbon of a neighboring nucleotide. Because two ester bonds form the link, it is called a phosphodiester bond. Ribonucleic acid (RNA) is produced when the sugar is a ribose, as shown above. When the sugar is deoxyribose, which lacks the –OH group on the number 2 carbon, the result is deoxyribonucleic acid (DNA). The nucleobases in the illustration are shown as a purine and a pyrimidine, without specifying which bases.

But instead of attempting to get the nucleotides to polymerize in an energetically uphill reaction, we can take a shortcut. This is what Leslie Orgel and his group at the Salk Institute described in the 1970s in a series of groundbreaking experiments. The idea was to remove a water molecule ahead of time by chemically activating the nucleotides. This was done by attaching imidazole to the phosphate group through an ester bond, and the imidazole becomes what is called a leaving group. If two such activated nucleotides come in contact with each other, the leaving group lives up to its name and leaves. The phosphate then spontaneously attaches to the neighboring ribose to make a phosphodiester bond in an energetically downhill reaction. In the next chapter, I will describe how Orgel took advantage of this reaction to demonstrate the first non-enzymatic replication of a nucleic acid, but here I want to show how Jim Ferris used activated nucleotides to make RNA polymers on clay mineral surfaces.

In the discussion above, I mentioned that for polymerization to occur, it is necessary to bring the reacting groups close enough together for a long enough time so that the reaction can occur. Jim Ferris, a chemist at Rensselaer Polytechnic Institute in Troy, New York, has explored the possibility that a clay mineral surface could serve as the organizing agent. He found that a clay mineral called montmorillonite, after a simple chemical treatment with an acid, could do precisely this. For instance, Jim can make an aqueous solution of adenosine monophosphate activated with the imidazole group. Nothing happens in the solution in the absence of clay, but as soon as the clay is added, the activated AMP binds to the mineral surfaces and begins to polymerize. This discovery fulfilled a prediction made 50 years ago by British chemist Desmond Bernal, who thought that clay might have such a catalytic property. With modifications of the method, Jim is now able to grow RNA molecules up to 50 nucleotides long, which would be long enough to fold into a ribozyme structure if just the right set of nucleotides were present.

This is a wonderful result, but there are several caveats. First is that the nucleotides need to be activated, and we don't know a plausible way to do this yet in a prebiotic scenario. Second is that RNA can form two types of bonds. In the biological RNA, the phosphate diester is called a 3'-5' bond because it links the oxygens on the third and fifth carbon of the ribose ring. However, in the absence of an enzyme-catalyzed synthesis, diester bonds also form between the second and fifth carbons. If a number of these 2'-5' bonds randomly occur in RNA molecules, there is no guarantee that the molecules will fold into a catalytically active structure.

Perhaps the most difficult sticking point involves chirality. Any ribose in the prebiotic mix is likely to be a racemic mixture of both D- and L-forms, but there is a major problem with such mixtures. Jerry Joyce demonstrated this point very clearly by trying to run Leslie Orgel's template-directed RNA synthesis using nucleotides containing both D- and L-ribose. The reaction was completely inhibited because the mixed nucleotides could not fit together on the template. Life today, of course, gets around these problems by using only the D-form of ribose or deoxyribose, and polymerase enzymes synthesize nucleic acids so that only 3'-5' phosphodiester bonds are produced.

CONNECTIONS

All life depends on a variety of polymers to function as structural components of cells, as enzymatic catalysts, and as carriers and transmitters of genetic information. The monomers of protein polymers are amino acids linked together through peptide bonds, and the monomers of nucleic acids are nucleotides composed of purine and pyrimidine bases, ribose or deoxyribose, and phosphate, linked together through phosphodiester bonds. For life to begin, there must have been some way for polymers to be produced spontaneously in the absence of enzymatically catalyzed reactions. This is not too difficult to imagine for simple peptides composed of a few kinds of amino acids that were likely to have been available in the prebiotic mix, but the spontaneous synthesis of RNA and DNA is a much more difficult proposition and represents a major gap in our knowledge.

Even when activated, polymerization does not necessarily proceed very far because the reactants are free to diffuse in solution and rarely undergo a collision that results in bond formation. This is where a mineral surface can play a role, and a clay mineral called montmorillonite has been shown to adsorb activated nucleotides and organize them in such a way that they undergo polymerization. It seems likely that mineral surfaces on early Earth also served to organize potential reactive monomers in this way.

11

CATALYSTS: LIFE IN THE FAST LANE

Life originated, according to the RNA World hypothesis, from self-replicating
ribozymes that catalyzed ligation of RNA fragments. We have solved the
2.6 angstrom crystal structure of a ligase ribozyme that catalyzes regiospecific
formation of a 5' to 3' phosphodiester bond between the 5'-triphosphate and the
3'-hydroxyl termini of two RNA fragments.

MICHAEL ROBERTSON AND WILLIAM SCOTT, 2007

Scientific breakthroughs can be very exciting for the people who make the discoveries.
Bill Scott is a faculty member in my department here at the University of California,
Santa Cruz, and when he and his graduate student Monika Martick began to get X-ray
diffraction patterns from crystals of ribozymes, Bill had this to say in an interview for a
news release in 2006:

> Monika e-mailed me from the Stanford Synchrotron at 3 a.m. to show me the most beauti-
> ful electron density map I had ever seen. I was so amazed I probably didn't sleep for the
> next three weeks.

Why can someone lose sleep over ribozymes? The reason is stated in the first quotation above from a later paper out of Bill's lab: It is highly probable that ribozymes were
central components of the first forms of life on Earth.

Ribozymes were first observed in 1983, and the startling discovery of an RNA catalyst won Nobel Prizes for Thomas Cech, now at the University of Colorado, and Sidney
Altman at Yale University. Their discovery led to an entirely new way to think about the
origins of life, to be discussed later in this chapter. But first we need to know something
about catalysts and why they are so important to life processes. The word "catalyst" was
originally a technical scientific term, but it has come into common usage to mean an activating factor. For instance, my daughter's soccer league is called Catalyst, and the league
has activated wonderful young girls who might otherwise be sitting in front of TVs. But
in chemistry there is a more specific meaning, which is that a catalyst is anything that

causes spontaneous reactions to proceed faster toward equilibrium. Furthermore, catalysts do not speed up all reactions, but instead are fairly specific. In order to understand the origins of life, we need to know what kinds of catalysts were available on early Earth and how they became incorporated into living systems.

BIOLOGICAL CATALYSTS

In Chapter 6, I discussed the changes in energy content that occur during a chemical reaction and showed how all spontaneous reactions move toward equilibrium at a certain rate. But a potentially reactive molecule faces a hurdle, a kind of hill called activation energy that must be added to the molecule in order for it to react. We also noted that the size of the hill influences the rate of a reaction, and that heat can add activation energy to reactants and therefore speed up a reaction. Catalysts work by reducing the size of the hill so that less activation energy is required. The catalyst transiently binds to one or more reactant molecules and alters the electronic structure in such a way that the reaction occurs more rapidly. Even though the catalyst binds to a reactant, the catalyst itself is not chemically changed during the reaction and can be used over and over.

Living systems long ago learned how to take advantage of catalyzed reactions. An important point is that many different substances can catalyze a reaction if they happen to be able to interact with a potential reactant. For instance, certain metals are catalysts, such as the platinum and palladium used in the catalytic converters of automobiles to reduce hydrocarbon emissions that contribute to smog. Other metals became incorporated in living systems because of their ability to act as catalysts. Iron, nickel, copper, manganese, and zinc are all essential trace elements required by living organisms. The reason is that they can form complexes with proteins that then are able to catalyze certain metabolic reactions.

It is reasonable to think that metals were present in the prebiotic environment and were already acting as catalysts in non-biological chemical reactions. The first forms of life discovered that some of those reactions were useful and incorporated them into a primitive metabolism, along with the metal cofactors. In order to understand how metals can contribute to such reactions as catalysts, we can begin by describing how enzymes catalyze a relatively simple reaction that involves oxygen, hydrogen peroxide, and water. Let's first consider the way that iron interacts with oxygen. If we expose iron to air, it readily combines with oxygen to become iron oxide, otherwise known as rust. The interaction of iron with oxygen is the key to many catalyzed reactions. For instance, when cells use oxygen to generate energy, toxic by-products called reactive oxygen species (ROS) are produced. One of these is hydrogen peroxide, which is dangerous because it can dissociate into hydroxide radicals:

$$HOOH \{\Rightarrow\} 2 \ ^{\cdot}OH$$

Recall from Chapter 4 that chemical bonds are produced by pairs of electrons between atoms. However, sometimes a molecule has an unpaired single electron, and these molecules are called radicals. The little dot on the OH in the above reaction signifies an unpaired electron, so this is a hydroxide radical. Unpaired electrons make radicals extremely reactive, which means that the hydroxide radical produced by dissociation of hydrogen peroxide can react with other normal components of a living cell, such as unsaturated lipid chains, proteins, and DNA, to produce molecular damage. Because of this reactivity, hydrogen peroxide is toxic.

But another reaction of hydrogen peroxide is dissociation into oxygen and water:

$$2\,HOOH \;\{\Rightarrow\}\; O_2 + 2\,H_2O$$

If this reaction can be speeded up, the chances of forming toxic hydroxide radicals are reduced. How can it be speeded up? By a catalyst, of course. Dilute hydrogen peroxide has antiseptic properties, and can be purchased as a 3% solution at a drugstore. If we poured some of this hydrogen peroxide into a glass and watched for several hours we would see a few bubbles appearing, which is the oxygen released spontaneously by the slow, non-catalyzed reaction. But if we added some iron ions, purchased as an iron sulfate nutritional supplement, we would see bubbles produced in minutes rather than hours. The iron atoms all by themselves act as catalysts. But then if we added iron in the form of an enzyme called catalase, which contains iron atoms surrounded by a protein structure, within seconds we would see the hydrogen peroxide release oxygen so fast that bubbles overflow the side of the glass. Catalase is one of the most efficient enzyme catalysts, and a single catalase can split millions of hydrogen peroxide molecules into water and oxygen per second. Catalase is present in most cells to protect against the toxic effects of hydrogen peroxide that would otherwise accumulate.

Iron and copper are present in virtually all enzymes and in some proteins that interact with oxygen, particularly the electron transport enzymes in mitochondria and hemoglobin, the oxygen-transporting protein in blood. However, as noted earlier in the book, there was virtually no oxygen in the atmosphere when life began. Was iron still a cofactor in enzymes? There was no oxygen, but there was plenty of sulfur, which is just below oxygen in the periodic table. Furthermore, iron reacts strongly with sulfur, just as it reacts with oxygen to produce iron oxide. A surprising discovery is that a number of enzymes contain iron-sulfur centers, some of which are also in the mitochondrial electron transport chain. Several researchers have proposed that iron-sulfur proteins evolved as catalysts in the earliest forms of life and have since been adopted by aerobic life.

To sum up, just as we assume that amino acids were available in the prebiotic environment, there must also have been a variety of compounds that could serve as primitive catalysts. The first such catalysts were likely to have involved metals like iron, copper, zinc, nickel, and magnesium that would have been present as ions in solution and also in the form of mineral surfaces. These metals are still deeply involved in life processes today, but their catalytic properties are greatly enhanced by binding to specific protein structures.

CATALYSTS GUIDE METABOLIC REACTIONS

Catalysts make reactions proceed faster toward equilibrium, but there is more to the story. Catalyzed reactions are highly specific, which means that even though numerous possible reactions might occur in a system, certain reactions will dominate if a catalyst is present because those reactions proceed fastest. This is how catalysts guide metabolic reactions. An example is the electron transport system of mitochondria, in which each step is guided by the presence of an enzyme. Mitochondria can be isolated from virtually any plant or animal tissue, and they are still functional for a few hours in the laboratory. If we add a substrate such as succinic acid to the mitochondria, an enzyme in the mitochondria called succinic dehydrogenase immediately begins to metabolize the substrate by removing electrons from its chemical structure. The electrons are transferred down a chain of enzymes, several of which contain iron. Like hemoglobin in blood, the bound iron gives the enzymes a distinct reddish color, and for this reason they are called cytochromes, from Greek words meaning "cell color." The electrons finally reach cytochrome oxidase, another iron-containing enzyme that specializes in donating the electrons to an oxygen molecule. Most of the oxygen you take in with every breath is used by cytochrome oxidase for this purpose.

But now let's do an experiment by adding carbon monoxide to the mitochondria. Something pretty amazing happens. The carbon monoxide strongly binds to the iron atom in the cytochrome oxidase and prevents oxygen from reaching the binding site. The result is that electron transport grinds to a halt, which is one of the reasons that carbon monoxide is toxic. Cyanide, by the way, would produce the same result if we added it to the system. The reason cyanide is so toxic is that it binds to the iron in hemoglobin of blood, again displacing oxygen. As a result, the blood color changes from the bright red of oxygenated hemoglobin to the darker purple of venous blood, which gives a characteristic bluish cast to the skin. In fact, the word "cyanide" is derived from the Greek word for "blue."

HOW DOES A PEPTIDE CHAIN BECOME A CATALYST?

Just because amino acids can form chains called proteins does not mean that the proteins automatically take on catalytic properties of enzymes. Instead, the chains must be composed of certain amino acids that are present in a specific sequence in the chain, and only then can the chain fold up into the structure we call an enzyme. Protein folding was discussed in Chapter 10 on polymers, with ribonuclease given as an example of a relatively simple protein catalyst. Now we can use a somewhat more complex example that illustrates how a metal atom is incorporated into the catalytic center of an enzyme molecule. The enzyme called carbonic anhydrase catalyzes the reaction by which carbonic acid dissociates into carbon dioxide and water:

$$H_2CO_3 \{\Rightarrow\} H_2O + CO_2$$

This is ordinarily a relatively slow reaction, which you can prove to yourself by opening a can of soda water. There is a pop when the can is opened and gas pressure is released, but after that, the carbon dioxide comes out over a period of hours depending on temperature. But as soon as you take a sip, the soda water foams up in your mouth. The reason is that small amounts of carbonic anhydrase are present in saliva and catalyze the reaction. Carbonic anhydrase is used by cells wherever carbon dioxide must be rapidly taken up or released. For instance, the chloroplasts of plants attach carbon dioxide to a sugar called ribulose in the first step of photosynthesis, and carbonic anhydrase increases the rate at which atmospheric CO_2 is made available for the reaction. Shellfish such as oysters form hard shells by reacting calcium with carbonate to produce calcium carbonate mineral, and carbonic anhydrase speeds the reaction by which carbon dioxide dissolved in sea water is changed into bicarbonate and carbonate. Another kind of carbonic anhydrase is present in red blood cells. Blood passing through capillaries in the lungs has only a limited time to release the carbon dioxide carried to the lungs from tissues, and carbonic anhydrase makes it possible for CO_2 to leave the red cells in a second or so.

Figure 24 shows the molecular structure of carbonic anhydrase, both as a space filling version on the left and as a ribbon structure on the right. Take a look at the ribbon structure, which illustrates how the string of amino acids has folded into the active enzyme molecule. The spirals are alpha helices, the ribbons are beta-sheet structures, and the

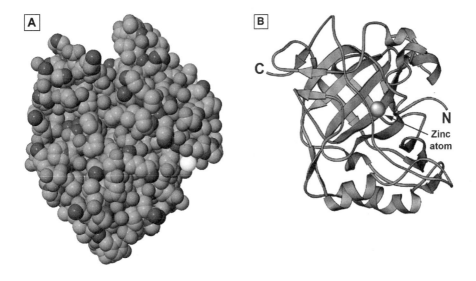

FIGURE 24
Carbonic anhydrase is illustrated as a space-filling model on the left, and as a ribbon structure on the right, which shows the essential zinc atom in the center. Many enzymes incorporate metal atoms within a protein matrix, which amplifies the natural catalytic activity of the metal. Other examples include the iron- and copper-containing enzymes that catalyze electron transport, and the molybdenum protein that is essential for the reaction in plants that strips electrons from water molecules to produce oxygen.

thin lines represent portions of the molecule that are neither helical nor sheetlike. Like most other enzymes that catalyze reactions of gas molecules, carbonic anhydrase incorporates a metal, in this case the zinc atom shown as a gray sphere in the center. The zinc is bound to certain amino acids in the folded structure of the protein and forms the active site of carbonic anhydrase. This is where carbonic acid (H_2CO_3) binds in the first step of the reaction. When it binds, the activation energy barrier is reduced, and carbon dioxide and water are released as products thousands of times more rapidly than in the absence of a catalyst.

WHAT CATALYSTS WERE REQUIRED FOR THE FIRST LIVING SYSTEM?

Life today uses enzymes to catalyze thousands of different reactions that fall into a dozen or so classifications. Instead of listing all the classes that have now been established, I will list only the ones that might have been useful in some way for the first forms of life (Table 4).

Most of the catalysts in the list are proteins composed of 20 different amino acids, and some have metal atoms embedded in the structure. An exception to this rule is catalytic RNA in the form of ribozymes, which act on themselves or other RNA molecules. Ribosomes are complex molecular machines composed of RNA and protein that catalyze the synthesis of the peptide bonds linking amino acids into proteins. As we will see later, the site deep in the ribosome where this happens is actually a ribozyme.

The fact is that there has been relatively little experimental work on catalysts required by the first living systems. For this reason, much of what I will say here is

TABLE 4   Primary Classes of Enzymes Related to Primitive Life

Enzyme	Reaction Catalyzed	Role in Cellular Function
Polymerases	Synthesis of ester bonds	Nucleic acids synthesis
Hydrolases	Hydrolysis of bonds	Digestion
Metabolism	Transformation	Preparing nutrients for use
Synthases	Synthesis of ATP	Energy production
Phosphorylases	Adding phosphate groups	Activation of metabolites
Electron transfer	Redox reactions	Energy production
Membrane transport	Permeation	Bringing nutrients into cells
Ribozymes	Bond breaking and splicing	RNA processing, protein synthesis
Ribosomes	Synthesis of peptide bonds	Protein synthesis

necessarily speculative, but this is how science works. First we speculate, guessing how something might work (a working hypothesis), and then we test those guesses experimentally. The way I think about the catalysts required by the first living system is to imagine a mixture of polymers encapsulated within a membranous compartment. One particular mixture happened to have sequences of monomers that give the polymers potential catalytic properties, so the question is how would that minimal set of catalysts give the system the property of growth and reproduction? Not to belabor the point, but once again, that is a fundamental question of biology and the central question addressed in this book.

The first essential process would be to capture certain kinds of nutrients and energy from the environment while excluding most other compounds that would not be useful. The nutrient compounds that must be brought into the cell include phosphate, amino acids, nucleobases, simple sugars, and essential trace elements. We don't yet know how membrane transport first appeared and then evolved to the complex systems used by modern cells, so let's just take one example and speculate on how it might have occurred. If we can figure out a possible mechanism for phosphate transport, we can perhaps apply it to other nutrients as well.

As always, we go to living cells today for inspiration. That brings us to the first revelation about membrane transport of nutrients, which is that most such processes require an input of energy. In the case of ion transport, the energy is supplied directly by ATP, which adds a phosphate to the pump protein to drive the process. In the case of phosphate, amino acids, and carbohydrates, the energy is in the form of ion gradients, and the transport process involves an enzymelike carrier in the membrane that transports both the nutrient and the ion, which are usually protons or sodium ions.

The bottom line is that the first cellular life needed some way to bring nutrients into the internal volume and then make them sufficiently concentrated so that they could undergo the biosynthetic reactions required for producing polymers. The only way to do that was either to use the chemical energy available in pyrophosphate bonds or to develop some sort of energy-dependent mechanism to produce ion gradients. Transport and concentration seem so fundamental to life that there must have been some mechanism by which energy-containing gradients were produced across the first cell membranes and some process by which the energy in those gradients could be used to transport nutrients into the cell. In my judgment, only protons could serve this purpose. The reason is that it is comparatively easy to produce a pH gradient of 3 pH units across a membrane, equivalent to a thousand-fold concentration gradient of protons, because only a few protons must be transported. For instance, to make a gradient from pH 8 to pH 5 we only need to go from 10 nanomolar protons to 10 micromolar protons. If a primitive cellular compartment was 2 micrometers in diameter, about the size of a typical bacterial cell, only six protons need be transported to produce a thousand-fold gradient. But if we tried to make the same magnitude gradient with sodium ions, we would need to pump 300,000 sodium ions to go from 0.5 molar (the concentration of sodium ions in sea water) to 0.5 millimolar. The calculated number of protons—six—assumes that no buffering is present, but bicarbonate (HCO_3^-) would be an abundant buffer in any water exposed to the early atmosphere, which was rich in carbon dioxide. Buffers react with protons, so the actual number of protons needed would be somewhat larger in a natural environment but still minuscule compared to the number of sodium ions.

The conclusion is that the relationship between living cells and protons is likely to have been part of life processes from the very beginning. But how could pH gradients be generated and then coupled to transport processes? Here is one scenario that can be tested experimentally. We know that early Earth was not at equilibrium, and therefore a variety of chemical reactions must have been continuously going on. Some of these would have been simple electron transfer reactions in which electrons on one species of atom or molecule called the donor are transferred to another set of atoms or molecules that are electron acceptors. An example I use when I teach biochemistry is the reaction between ascorbic acid (vitamin C) and a dye called methylene blue. When I mix them together, ascorbic acid donates electrons to methylene blue, and the solution changes from a deep blue color to clear because the reduced form of the dye does not absorb light. Then I add ferricyanide, which is simply a soluble form of iron with six cyanide molecules surrounding it. The reduced methylene blue gives its electrons to the iron in the ferricyanide and the solution becomes blue again.

The blue dye has its electrons present in the chemical bond between oxygen and hydrogen (−O:H), and this leads to the most important point to be made here. When the electrons are given to iron, the hydrogen comes off in solution as H+, and the solution becomes acidic. Why is this important? The reason is that we can carry out the reaction in such a way that the donor is separated from the acceptor by a membrane. Years ago, we showed that it was possible to trap ferricyanide in lipid vesicles with ascorbic acid on the outside. Nothing happens because the donor has no way to react with the acceptor separated by a membrane. But if we then add any of several dyes that are permeable to the lipid bilayer, something amazing happens. Electrons are transferred from the ascorbic acid to the dye, which also takes up hydrogen ions from the water when it becomes reduced. The dye readily permeates the bilayer and gives the electrons to the ferricyanide inside, simultaneously releasing the hydrogen ions. As a result, a pH gradient as large as 4 pH units can be produced, more than enough energy to synthesize ATP if an ATP synthase was embedded in the membrane. When Peter Mitchell first proposed chemiosmosis, he included just this kind of electron transport linked to proton transport as a way to pump up proton gradients across membranes, and we now know that a quinone called ubiquinone functions this way in micochondria, and plastiquinone does the same in chloroplasts.

This is just a model reaction carried out under laboratory conditions, but there must have been similar reactions possible with compounds in the environment of early Earth as soon as membranous vesicles self-assembled. For instance, aromatic compounds like naphthol can undergo oxidation-reduction reactions. Such compounds are abundant components of meteoritic organics and must have been widely distributed in the organic mix on prebiotic Earth. This is an open question for future research. If we can find a plausible prebiotic reaction that produces pH gradients, we will be able to conclude that chemiosmotic mechanisms could make the energy of proton gradients available to the earliest forms of life.

Now comes the more difficult question: How could the energy available in a pH gradient be coupled to a transport process? One possible answer is that a membrane-associated polymer such as a peptide could catalyze a mechanism in which the transport of a proton down the pH gradient was coupled to the inward transport of nutrients such as phosphate or amino acids. Because life today could not exist in the absence of energized nutrient transport, it is reasonable to think that such processes were also essential for the origin of cellular life. No one has yet begun to study coupled transport in the context of the origin of life, so this question is wide open for future investigations.

ORIGIN OF METABOLIC ENZYMES

Four billion years ago, the entire planet was a vast experiment in chemistry and physics. Chemical reactions would be occurring wherever energy impinged on mixtures of potential reactants, driving the synthesis of ever more complex molecules. In earlier chapters, we described a number of these reactions, but let's just review a few of the more important ones here. In the atmosphere, light energy and electrical energy were producing compounds like formaldehyde and cyanide, which could in turn react to produce molecules like amino acids, nucleobases, and sugars that would dissolve in bodies of water. Wherever dilute solutions of these compounds were heated and dried or activated by exposure to activating agents like COS, condensation reactions occurred to produce an enormous variety of different polymers. It is virtually certain that certain components of this chaotic mass of reactions became entrained in pathways leading to the earliest forms of life, but how could this happen? The quick answer is that we don't yet know, but that should not keep us from doing what science does best, which is to make guesses called hypotheses and then test them in the laboratory. I will describe several guesses here, and then in Chapter 14 I will reveal my own guess.

Several of the guesses are known metaphorically as "worlds." I will mention a few of these here and the researchers who invented them, and then show how catalysts were involved in each. The use of the word "world" in this context simply means that a given reaction pathway or process was pervasive on early Earth and represented either a prebiotic era that preceded the origin of life or a simpler version of life that evolved into the DNA-RNA-protein world of life today. These worlds are summarized in Table 5, and each focuses on a set of primary components such as coacervates, proteinoids, clays, sugars, iron-sulfur minerals, thioesters, lipids, and RNA. These are described elsewhere

TABLE 5   Catalytic/Organizing Components in Proposed Scenarios for the Origin of Life

Primary Component	Proposed by	Properties
Coacervates	Aleksandr Oparin	Self-organizing polymeric structures
Proteinoids	Sydney Fox	Self-organizing polymers of amino acids
Clay surfaces	Graham Cairns-Smith, James Ferris	Clay surfaces catalyze synthesis of polymers
Sugar world	Arthur Weber	Simple molecules contain chemical energy that drives primitive metabolism
Geochemical catalysts	George Cody, Robert Hazen	Mineral interactions with organic compounds promote primitive metabolic reactions
Iron-sulfur world	Gunther Wächtershäuser	Minerals composed of iron, nickel and sulfur catalyze metabolic reactions
Thioesters	Christian DeDuve	Sulfur bonds contain chemical energy to drive metabolic reactions
RNA world	Leslie Orgel, Francis Crick, Gerald Joyce, Walter Gilbert	RNA in the form of ribozymes serve both as catalysts and genes
Lipid world	Soron Lancet, Daniel Segre	Organized lipid structures have catalytic properties and contain information in their composition

in the book, so the point I am making here is that each of these worlds has the potential to contain a series of catalyzed reactions that direct the reactions in a way that would be useful to the first forms of life.

I will choose just one of these to illustrate a series of reactions that have the potential to evolve into a primitive metabolic pathway with catalyzed reactions. This was first proposed by Harold Morowitz at George Mason University, who noticed that there are patterns within metabolic pathways that make sense if one thinks of them as arising from a relatively simple fundamental pathway of glycolysis. Furthermore, the patterns make sense in terms of thermodynamics. That is, they can be driven by the free energy available in a specific set of chemical reactions. For instance, amino acids fall into four classes, each of which stems from one of the intermediates in glycolysis by addition of ammonia (NH_3). Furthermore, with few exceptions, the codons of each of the four families begin with a different base: cytosine for the a-ketoglutarate family, adenine for the oxaloacetate family, uracil for the Embden-Meyerhof family, and guanine for a fourth family of exceptions. Morowitz suggests that this is too striking a pattern to be mere coincidence. We can speculate that when the most primitive versions of metabolism started up in the membranous compartments of the first cellular life, cells that contained small proteins capable of catalyzing the addition of NH_3 to a glycolytic intermediate would have a distinct advantage over cells that lacked these proteins. Furthermore, cells that could incorporate the information in a genetic code specific for each amino acid family would have an advantage over those cells that lacked such a code. How this actually transpired is an open question for future researchers delving into the metabolic aspects of life's first catalyzed pathways.

RIBOZYMES AND THE RNA WORLD: THE ORIGIN OF POLYMERASES

Isaac Asimov is considered to be one of the great science fiction writers of our time. Several of his novels, including *I, Robot,* have inspired movie versions. Less well known is that Asimov was trained as a biochemist and had the title professor at Boston University, although he spent his time writing, rather than leading the traditional academic life. One of Asimov's quotations that has always stuck with me is the following: "The most exciting phrase to hear in science, the one that heralds new discoveries, is not 'Eureka!', but 'That's funny . . .'" In the late 1970s, Tom Cech was a young professor of molecular biology at the University of Colorado, Boulder, and was studying a special catalyzed process in protozoa called Tetrahymena. When messenger RNA is synthesized from DNA it often has one or more sections in the chain called introns. The origin of introns is still being debated, but one idea is that they represent older DNA sequences that are no longer used yet are still inherited. What is clear is that introns must be removed before the messenger RNA can be translated into a functional protein by ribosomes. This reaction is called splicing and involves the cutting of the strand at the beginning and end of the intron and then attaching the two ends to make mRNA that can be used by the ribosome. To study splicing in the Tetrahymena RNA, it was first necessary to isolate the RNA. The problem was that no matter how hard Cech and his students tried, the RNA they isolated seemed to go through the splicing reaction all by itself, even in the complete absence of any contaminating protein enzymes. Asimov would have said, "That's funny . . ."

At about the same time, another young professor, Sid Altman at Yale University, was studying an enzyme called RNAse P, which was involved in processing molecules of transfer RNA. It was known that RNAse P in bacteria was composed of both protein and RNA, so Altman decided to take it apart to study the properties of each component. But when the RNA component was added to the tRNA, it could catalyze the reaction all by itself. That was funny too.

In the 1970s, there was no evidence whatsoever that a nucleic acid could have an active catalytic site, although as early as 1969, Carl Woese, Leslie Orgel and Francis

Crick suggested that this might be possible. The accumulating evidence forced Cech and Altman to propose the same basic idea in two separate papers in 1983. The proof that RNA could act as a catalyst was revolutionary, and Cech and Altman were awarded Nobel Prizes in 1989.

Meanwhile, Harry Noller and his research team at the University of California, Santa Cruz, were using a variety of techniques to study ribosome functions and particularly how the active site on the ribosome could add an amino acid to a growing peptide chain during protein synthesis. This process is the peptidyl transfer reaction discussed in Chapter 10. The way to approach such problems is to simplify the structure, so Noller and his research team began to use various methods to strip away the proteins of the ribosomes while they measured the ability to catalyze the peptidyl transfer reaction. Surprisingly, most of the proteins could be removed, but the reaction persisted. Then Noller's team crystallized the greatly simplified structure and performed X-ray diffraction analysis of its molecular structure. The active site turned out to be associated with one of the ribosomal RNA components called 23S, a somewhat cryptic abbreviation left over from early separations using high-speed centrifugation. The crystal structure showed that the small amount of remaining protein was too far from the active site to play a role in the catalyzed reaction. There could be only one conclusion, which is now generally accepted: The active site of peptidyl transfer in protein synthesis is a ribozyme!

The fact that RNA could take the form of catalytic ribozymes inspired Walter Gilbert at Harvard University to publish a one-page letter in *Nature* entitled "RNA World" that seemed to resolve the chicken-and-egg question of what came first: proteins or nucleic acids. Gilbert reasoned that a primitive version of life might use RNA both as a catalyst and as a carrier of genetic information, which then evolved into a more efficient version in which DNA stored genetic information and proteins became the primary catalysts.

Twenty-five years later, we know that ribozymes have catalytic roles in a surprising number of biological functions. Many of them also require metals such as magnesium and zinc ions for their activity. There are even infectious versions of ribozymes called viroids. These are the smallest known infectious particles, containing only 200 nucleotides in their complex folded structure, and are responsible for certain diseases of plants. When viroids enter a plant cell, they initiate a process that uses polymerases in the cell to synthesize more viroids, ultimately resulting in cell death.

Ribozymes are nucleic acid polymers, so they are readily synthesized by polymerase enzymes in the laboratory. This is a much more convenient process to study than protein synthesis, which requires ribosomes and 100 or more other components. Several research groups have taken advantage of this fact to evolve synthetic novel ribozymes, as will be described in the next chapter. Most known ribozymes have been crystallized and studied by X-ray diffraction. In general, ribozymes turn out to be hairpins or the hammerhead structure shown in Figure 25.

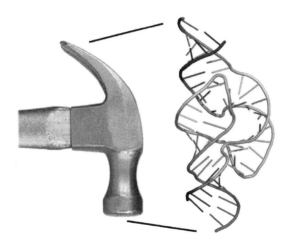

FIGURE 25

A ribozyme is an RNA molecule that catalyzes a chemical reaction. Many natural ribozymes catalyze either the hydrolysis of one of their own phosphodiester bonds, or the hydrolysis of bonds in other RNAs. One of the more common ribozymes is referred to as the hammerhead, and one such structure is shown here. It was recently discovered that the active site in a ribosome is also a ribozyme that catalyzes the peptidyl transferase reaction of protein synthesis.

The central property of life today is growth of polymers from monomers, so it seems reasonable to think that something like a polymerase was an essential catalytic activity for life to begin. Perhaps most significant for origins of life research are attempts to evolve a ribozyme that can act as a polymerase and catalyze its own replication. The initial research was carried out by Jack Szostak and Jennifer Doudna at Harvard Medical School, and they were later followed up by David Bartel at the Whitehead Institute, who was able to produce a ribozyme that could copy up to 14 nucleotides in a sequence that was part of its structure. More recently, Peter Unrau at Simon Fraser University in Canada has improved this to a record 20 bases.

CONNECTIONS

Most biochemical reactions that are the foundations for life today are speeded up by catalysts called enzymes, which are proteins with sequences of amino acids that cause them to fold into highly specific tertiary structures. When several amino acids are grouped together in the folded structure, they form an active site where the catalyzed reaction takes place. The active site first binds the reactants and then interacts with their electronic structure in such a way that the activation energy is reduced, which increases the rate at which the reaction proceeds toward equilibrium. The reactions in a metabolic system are also guided by catalysts because enzymes have remarkable selectivity in regard to the structure of the substrates they use. The catalyzed reactions are speeded up compared to potential competing reactions, so the flow of metabolism is guided along those pathways having the fastest rates.

The first forms of life used certain chemical reactions that could be speeded by catalyzed reactions, so there must have been catalysts existing in the prebiotic environment that could be incorporated as catalysts. Many enzymes today require metal atoms like iron, nickel, and zinc in their structures, and magnesium is often required as well. Metals having weak catalytic activities would have been abundant components in the prebiotic environment, and it seems likely that the first forms of cellular life made use of this by embedding such metals in peptides.

Most biological catalysts are proteins, but a few are composed of RNA and are called ribozymes. Because ribozymes can act as catalysts and also store genetic information, it has been proposed that life began as an "RNA World." RNA-based life then evolved into a biological system that uses three different molecular species for these separate functions: catalytic proteins, genetic information stored in DNA, and RNA as a carrier of genetic information from DNA to sites of protein synthesis.

COPYING LIFE'S BLUEPRINTS

Populations of various cross-replicating enzymes were constructed and allowed
to compete for a common pool of substrates, during which recombinant
replicators arose and grew to dominate the populations. These replicating
RNA enzymes can serve as an experimental model of a genetic system.

TRACY LINCOLN AND GERALD JOYCE, *SCIENCE*, 2009

It is common knowledge now that DNA is a double helix, and we know so much about
the function and structure of DNA that we tend to forget that there must have been a
much simpler version of a polymer that was able to grow and then replicate in some way.
"Replication" is an amazing word. It captures the essence of life, distilling it down to the
molecular structure of DNA. The quotation above is taken from a paper that demonstrates how close we are to understanding how a replicating system of ribozyme catalysts
can function. So in this chapter we will first describe the way that replication occurs in
all life today, and then we will ask "why" questions related to the origins of life: Why is
a double helix required for replication? Why are nucleotides the monomers of nucleic
acids? Why does life use nucleotide triphosphates to make DNA and RNA, and what
were the first nucleotides? But to understand what this means, how it occurs today, and
how it might have occurred in the first forms of life, we need to know about several fundamental properties of nucleic acids. So let's begin with a history lesson.

HOW DOES REPLICATION OCCUR NOW?

Replication is basically a polymerization process that requires a template and a primer,
and the product of the reaction is a new strand of DNA that is complementary to the
template strand. The idea of a template can be a little mysterious, so I will begin by
describing a simpler polymerization reaction that synthesizes a nucleic acid in the absence of a template. This contrast will clearly demonstrate what we mean by a template
and primer, and why life depends on them. In 1955, with the structure and significance

of DNA available from Watson and Crick's paper, the race was on to find enzymes that could synthesize nucleic acids. It had earlier been shown that if radioactive nucleoside diphosphates (NDPs) were added to bacteria that had been broken up to release their contents, the NDPs disappeared, and a large amount of radioactive RNA was produced. In 1955, Marianne Grunberg-Manago and Severo Ochoa at New York University isolated an enzyme called polynucleotide phosphorylase that could make RNA by using the energy of a pyrophosphate bond—but not the usual nucleoside triphosphates. Instead, the enzyme used nucleoside diphosphate in the following reaction:

$$(RNA)_n + NDP \{\Rightarrow\} (RNA)_{n+1} + P_i$$

The $(RNA)_n$ is called a primer because the enzyme needed a short piece of RNA to start the process. However, there were several strange properties of the enzyme. It could use any or all of the four NDPs as substrates, and it did not require a template. It became clear that this was not the enzyme responsible for replication, but rather its usual function in the cell, which is to break down messenger RNA (a polynucleotide) into its subunits. The enzyme does this by adding phosphate to the ester bond at one end of an RNA molecule, a reaction called phosphorylation, so that the nucleotide at the end of the chain falls off. The enzyme then proceeds to the next nucleotide, and the next, until the entire chain has been broken down into its monomers in the form of NDPs.

$$RNA + phosphate + PNPase \{\Rightarrow\} nucleoside\ diphosphates$$

PNPase is a beautiful ring structure composed of three subunits. A single strand of the RNA passes through the center of the ring, where each subunit takes part in adding phosphate and breaking off the nucleotide at the end in the form of the NDP. The main point is that this reaction is readily reversible. If NDPs are present in high concentration, along with a short chain of RNA, the enzyme turns right around and begins to synthesize RNA by adding nucleotides to the chain. But there is no template, so the nucleotides in the RNA reflect the composition of the NDPs. If only uridine diphosphate (UDP) is present, the RNA is composed entirely of uracil as the base, and the RNA is called poly-uridylic acid. If all four NDPs are present, the resulting RNA is a random mix of all four in the chain.

We can now compare this reaction to replication as it occurs in living cells, in which a template is used, and then see if we can establish a testable working hypothesis for how it might have worked in the first cells as life began. Watson and Crick hinted at replication in their original 1953 paper describing the double helical structure of DNA, and they answered the first why question above. DNA is a double helix because that is the only known way for a molecule to transmit a copy of itself between generations. The double helix can be split into two single strands, and then each strand becomes a template for the synthesis of a new double helix that contains all of the original sequence information, which is then given to each of the two daughter cells during cell division. In 1957, Crick published a landmark review in which he coined the term "Central Dogma" to describe how DNA could not only carry genetic information in its base sequence but could also use it to direct protein synthesis. The genetic information in one of the strands of the double helix is transcribed into messenger RNA, which, in turn, is used by ribosomes to synthesize proteins. There are 20 smaller RNA molecules called transfer RNA, each of which binds a specific amino acid with a high-energy bond and then carries it to a ribosome where it is added to the growing peptide chain.

But this means that there must be a code that translates a sequence of bases in the nucleic acid into the addition of a specific amino acid. Remember George Gamow from Chapter 1, who proposed that everything began with a Big Bang? Gamow was also intrigued by the question of a genetic code, and by pure force of logic, he came up with a proposal in 1954 that turned out to be correct. Gamow knew that genetic information could not be a singlet code in which one base in the nucleic acid codes for one amino acid. Nucleic acids are composed of just four bases, but there are 20 amino acids that need codes. For the same reason, it could not be a doublet code because if you do the arithmetic, there are only 16 ways in which four bases can be combined as doublets—still not enough for 20 amino acids. Therefore, Gamow proposed a genetic code composed of three bases, or triplets, which allows 64 combinations, more than enough for 20 amino acids. Each triplet, called a codon in the mRNA that is transcribed from the DNA, codes for a specific amino acid. Only tryptophan (UGG) and methionine (AUG) have unique codons, and AUG is also used as a "start" codon, which tells the ribosome where to start translating the base sequence in mRNA into a protein. There are three "stop" codons, one of which is UGA, that don't code for an amino acid, but instead tell the ribosome to stop the translation process. In the 1960s, Gobind Khorana at the University of Wisconsin, Robert Holley at Cornell University, and Marshall Nirenberg at the National Institutes of Health took on the task of establishing the genetic code for each amino acid and found that it really was a triplet code. Now all that was needed was an enzyme to make replication occur.

While all this was happening, Arthur Kornberg was a professor at Washington University in St. Louis, with an abiding interest in how DNA was synthesized by cells. Starting with extracts obtained by grinding up the common bacterium *Escherichia coli*, Kornberg discovered that he could add a mixture of the four nucleotide monomers of DNA as their triphosphates, along with a known amount of DNA. When the amount of DNA was measured later, there was more present than at the start of the experiment. Starting in 1956, Kornberg and his research group, which included his wife, published a series of papers that finally culminated in a purification of an enzyme now referred to as DNA polymerase I. The significance of this achievement did not go unrecognized, and Kornberg was awarded the Nobel Prize in 1959.

Although this was the first polymerase to be identified, it turned out to make a relatively minor contribution to nucleic acid synthesis in *E. coli*. DNA polymerase I was relatively slow, adding about 20 nucleotides per second, and in 1970, Kornberg's son Thomas discovered another enzyme called DNA polymerase III that does most of the work at nearly 1,000 bases per second. The actual process of DNA replication is complicated by the fact that both strands need to be duplicated, and in bacteria, four different polymerase enzymes are required just for synthesis and proofreading, along with other proteins that regulate the process.

HOW DO POLYMERASES REPLICATE DNA?

The answer seems simple: A polymerase grabs a single strand of DNA and uses it as a template. But the story is complicated by the double helical structure of DNA and by the way that monomers called nucleotides are added to the two chains by the polymerase. The word "nucleotide" sounds pretty intimidating if you have not had a college course in biochemistry, but in fact, nucleotides are not at all hard to understand. A nucleotide has three simple components: a base, a sugar, and a phosphate. The sugar is either a ribose in RNA or a deoxyribose in DNA, and there are only five bases to keep track of: ATGC (adenine, thymine, guanine, and cytosine) for DNA, and AUGC (adenine, uracil, guanine, and cytosine) for RNA. The truly astonishing thing is that these bases are the alphabet of life: Just four bases in DNA molecules can spell out the information needed to produce a bacterium, a mouse, or a human being—and that information is encoded in the sequence of bases in the DNA present in every living cell. The DNA molecule in a bacterial cell contains five million base pairs and is about 1 millimeter long. The DNA in the 46 chromosomes of a human cell has three billion base pairs and is 1,000 times longer, a little over 1 meter if laid out end to end.

I used the term "base pair" above, which is the secret of replication. In 1951, when James Watson and Francis Crick were just beginning to think about the molecular structure of DNA, it was already known that the amount of A in DNA was always equal to the amount of T, and G always equaled C. This was a deep clue to the structure of DNA and the mechanism by which it could be enzymatically replicated, but it still took two years of work to understand the significance. Linus Pauling, who first established hydrogen bonding as a primary stabilizing force in protein structure, published his best guess and got it wrong. Watson and Crick reserved their understanding of its significance for the famous last sentence of their paper in 1953: "It has not escaped our attention that the specific pairing we have postulated immediately suggests a possible copying mechanism for the genetic material."

What Watson and Crick finally realized is that the structure of adenine allows it to form two hydrogen bonds with thymine, and guanine can form three hydrogen bonds with cytosine, as shown in Figure 22 in Chapter 10. These are now called Watson-Crick base pairs, or complementary base pairing. The hydrogen bonds are strong enough to hold together the double helix of DNA, but not so strong that they cannot be pulled apart when the cell needs to replicate DNA or transcribe the nucleotide sequence of a gene to messenger RNA during protein synthesis. The relatively weak hydrogen bonds between bases leads to an important property of the double helix, which is that simple heating causes the double helix to separate into two single strands. This is called melting, which, depending on chain length, begins at around 60°C and is complete at 80°C to 90°C. Melting of DNA has little to do with life today, but needs to be considered when we think about how life may have begun in a hot environment of early Earth. It is also a basic step in polymerase chain reactions (PCRs), an essential method for amplifying DNA that will be described later.

The other property of the DNA double helix is that the two strands are not mirror images of each other, but instead have a direction that is specified by the bond between the sugar and phosphate. In order for biochemists to keep track of which bonds they are talking about, the carbons in the deoxyribose molecule are numbered from 1 to 5. In one strand of the double helix, the phosphodiester bond links the number 3 carbon of one deoxyribose to the number 5 carbon of the next, and this is called the 3'-5' direction. The other strand has the bonds in the 5'-3' direction. In general, these are referred to as the parallel and antiparallel strands. The reason this is important is that the enzymes that transcribe or replicate DNA can only read the nucleotide sequence in one direction. When the strands are being replicated, the DNA polymerase must read in the 3'-5' direction, and when DNA is being transcribed to RNA, one of the strands, the template strand, is also read in the 3'-5' direction, so the resulting messenger RNA is therefore synthesized in the 5'-3' direction. When the mRNA gets to the ribosome, it is the 5' end that initiates protein synthesis.

During DNA replication, *both* strands must be replicated at the same time, starting at the same place, but the polymerase can only read in the 3'-5' direction. This means that DNA replication in a living cell must be a pretty complicated process! I will give a brief sketch of the overall process in order to compare it to possible non-enzymatic replication that provided a way for the first forms of life to pass genetic information between generations.

THE NUTS AND BOLTS OF REPLICATION

As with every biosynthetic process where chemical bonds must be synthesized, the polymerization reaction of replication needs an energy source. Chapter 6 described how adenosine triphosphate (ATP) is the energy currency that drives most such reactions. But adenine is also one of the bases of DNA, so how can ATP provide energy for the other bases that need to be added? What happens is that there is a set of enzymes that allows ATP to give its chemical energy to the nucleotides of guanine, cytosine, and thymine (G, C and T) so that these are all present in cells as their high-energy triphosphate forms. These are the activated substrates used by polymerases to replicate DNA.

After protein synthesis, DNA replication is one of the most complicated enzyme-catalyzed reactions that go on in living cells. The description to follow is for bacterial replication, but a similar process occurs in eukaryotic cells using a different set of enzymes to accomplish the same end. DNA is present as a double helix with the two strands running in opposite

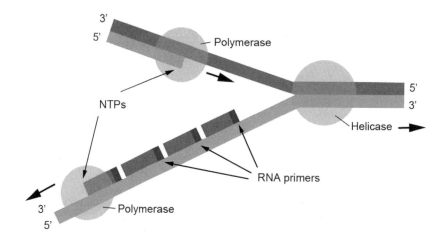

FIGURE 26

Replication of DNA by a DNA polymerase. Living systems require a double helix molecule in order to reproduce genetic information that is transferred between generations. During replication, an enzyme called a helicase unwinds the double helix, and each strand then serves as a template that is duplicated by the action of a polymerase enzyme. One of the template strands is called the leading strand and is directly replicated by DNA polymerase III using nucleoside triphosphates (NTPs) as activated monomers. The other strand, or lagging strand, must be replicated by a more complicated process which is described in the text.

directions, but the enzymes that replicate DNA need single strands to act as a template. This is solved by enzymes called topoisomerases and helicases, which latch onto the double helix of DNA at specific points and use the energy of ATP hydrolysis to unwind and split the helix into transient single strands—one with its base sequence going in the 3'-5' direction and the other with its base sequence going in the 5'-3' direction (Figure 26). DNA polymerase I binds to the 3'-5' strand and begins to synthesize a continuous complementary strand called the leading strand. What actually happens is that the enzyme wraps around the single strand with an active site that resembles a right hand with a thumb and fingers. A single nucleotide in the site, for instance a G, is exposed and needs a cytidine triphosphate (CTP) as a complementary base to be added. All four bases are present in the cytoplasm as nucleoside triphosphates, and these diffuse in and out of the site thousands of times per second until a CTP happens to hit the site in just the right way so that it binds to the active site, forming hydrogen bonds with a guanine base in the template strand. The active site catalyzes the loss of a pyrophosphate as a leaving group, leaving cytidine monophosphate (CMP) behind, which then forms an ester bond with the deoxyribose at the end of the growing chain. The enzyme then moves on to the next base in the DNA strand, which might be a T, and waits for an ATP to arrive and repeat the cycle.

But what about the other strand? This is where it gets complicated because the template strand to be replicated runs in the 3'-5' direction, the wrong direction for the polymerase. The solution is that an enzyme called a primase adds a short RNA primer near the split, and another DNA polymerase uses the primer to begin replicating the lagging strand in the 5'-3' direction, the opposite direction to the other polymerase synthesizing the leading strand. (This enzyme, by the way, is the DNA polymerase I that was discovered by Arthur Kornberg.) Replication of the lagging strand can only go on for a limited number of bases because the DNA is a helix that is being unwound as replication proceeds. In bacteria, sequences about 1,000 bases long are replicated before the polymerase falls off, which means that the lagging strand is present as short segments with the RNA primers still attached. These were discovered in 1968 by Reiji Okazaki and his wife Tsuneko at Nagoya University and are now called Okazaki fragments. In order to produce a continuous lagging strand, an enzyme called endonuclease removes the RNA primer and the gap is filled in with the correct bases by DNA polymerase I and then a second enzyme called a ligase attaches the ends of the Okazaki fragments together to make a complete copy. The result of all this is that the original double helix has become two identical double helices.

Despite the complexity of replication, the rate at which all this occurs is amazing! Consider that an *E. coli* cell can reproduce in about an hour, which means that it must replicate its entire genome of five million base pairs of DNA at 1,000 bases per second. It is also astonishing to realize how precise replication must be. There are approximately 100 trillion cells in the adult human body, each having 23 pairs of chromosomes with a total of three billion bases in DNA molecules, which would be over 1 meter in length if lined up end to end. The cells all originate in a single cell, the fertilized ovum, which multiplies in geometric progression (1, 2, 4, 8, 16 . . . and so on) around 50 times. At each division, the entire DNA of the genome must be precisely replicated, and even though different genes are turned on and off to produce different tissues, the base sequence remains unchanged. Another way to get a sense of the incredible precision is to consider that occasionally, a fertilized ovum happens to become separated into two cells after the first division. Each cell develops into a separate fetus. The genetic programming is so precise that the end result is a pair of human babies being born, twins who are virtually identical in every way.

THE POLYMERASE CHAIN REACTION: CYCLES OF REPLICATION

In the early 1980s, Kary Mullis accepted a job at Cetus Corporation in Emoryville, California, to work on the synthesis of short strands of nucleic acids called oligonucleotides. He was well prepared, with an undergraduate degree in chemistry from Georgia Tech, and a PhD in protein biochemistry from UC Berkeley. A few years earlier, Mullis had decided to drop out of science and write novels, even working for a couple of years at a bakery in Kansas. Then he followed the advice of a friend who suggested that he get back into science, first as a post-doc at UC San Francisco and then as a laboratory scientist at Cetus.

In the spring of 1983, as described in his biographical notes, Mullis was on a nighttime drive with his girlfriend into the California mountains to a cabin he was building when he got an idea. He knew three things that all had to work together to make the idea work:

- DNA polymerase can replicate a single strand of DNA.
- DNA polymerase requires a short primer to initiate the replication process.
- Double-stranded DNA can be melted at elevated temperatures into two single strands, and it can be annealed at lower temperatures.

The gist of Mullis' PCR idea is that two short primers can be added to DNA, one for each strand of the DNA after it is melted, along with a DNA polymerase. After the chains separated, the mixture could be cooled down so that the primers would anneal to their complementary sequence on each strand. Then the DNA polymerase would bind to the primer sites and replicate both strands in the 5'-3' direction, effectively doubling the amount of DNA present (Figure 27). But don't stop there. Heat up the mixture again to melt the double-stranded products and repeat the process, ending up with four times the original amount of DNA. If you do the arithmetic, you will see that after only 10 such cycles there will be 1,000 times more DNA than at the start, and after 20 cycles the original DNA will have been amplified a million-fold.

Mullis hurried back to Cetus to try it out, using a DNA polymerase isolated from *E. coli*. On December 16, 1983, it worked! But there was a problem. After every heat cycle, the polymerase was denatured, and more had to be added, which gets very expensive. Despite this limitation, Cetus recognized the potential value of PCR and assigned

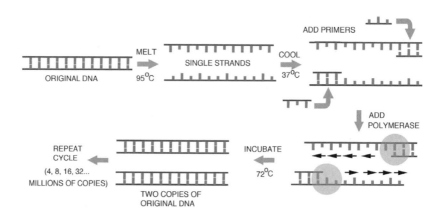

FIGURE 27

The polymerase chain reaction (PCR) depends on melting of a DNA double helix into single strands, followed by binding of primers and action of a polymerase enzyme. In the diagram, the primers are shown binding to single strands after a heat treatment melts the original double helix, followed by replication of both strands by the Taq polymerase. After the cycle is repeated 40 or more times, the DNA is amplified so that a few nanograms of starting material become micrograms of product, which is sufficient for modern analytical methods.

other scientists to develop some applications, and took out patents on the process. Mullis continued to work on improving the method, and in 1986 realized that he could use a polymerase isolated from a thermophilic bacterium named *Thermus aquaticus*. This enzyme is now called the Taq polymerase, an abbreviation of the bacterial source, and it easily withstands the near boiling temperatures required to melt long strands of DNA. The result was a revolutionary method to amplify minuscule amounts of DNA, even the tiny fragments preserved in the teeth of Neanderthals 30,000 years old from which we now have a complete genome.

The idea Kary Mullis had while driving to his mountain cabin ultimately was worth $300 million to the Cetus Corporation, which sold the patent to Roche Pharmaceuticals. Kary Mullis only received a $10,000 bonus from Cetus, and he left the company in 1986 to work with another company, Xytronyx in San Diego. Mullis shared a Nobel Prize in 1993, and most recently founded his own company to develop an idea he has for promoting a rapid immune response to viral infections.

Now let's see how PCR actually works because it is possible that a natural version of PCR might have given rise to the first replicating system of molecules. Figure 27 shows what happens during the first two cycles. This assumes that the target DNA has already been purified, usually isolating a single band off an electrophoretic gel. The first step is to melt the double helix of the target DNA for a minute at 95°C in the presence of two primers that have complementary sequences to the target DNA. The solution is then cooled for a minute to allow the primers to anneal, and then it is warmed to the optimal temperature for the Taq polymerase to work. The polymerase has already bound to the primer sites on the two separated strands of DNA, and it begins to replicate the DNA in the 5'-3' direction on both strands. This works at a rate of around 1,000 bases per minute, so in 2 minutes, the reaction is completed, and the second heating cycle can begin. A typical run might include 30 to 40 such cycles and could go on indefinitely, except that the polymerase exhausts the supply of NTPs.

Could something like this work in the prebiotic environment? There were no polymerases as such, but the temperatures were in the range required for PCR. When I visit geothermal sites on the flanks of active volcanoes, it is not hard to find hot springs with water temperatures of 90°C and above, while around the borders of the hot springs, the temperature of the particulate minerals—clays and lava surfaces—are quite a bit cooler, by tens of degrees Celsius. Furthermore, hot springs don't just sit there and simmer. Instead, there is likely to be active boiling, even geyser-like activity that splashes the hot water onto the warm surrounding rocks. The range of 20°C to 30°C between hot water and warm rocks is very close to the temperature range used in PCR.

The fact of active splashing and evaporation brings us to the second cycle that I think is important. Organic compounds are relatively dilute in the hot water, but when they dry, they become much more concentrated, forming thin films on the surrounding mineral surfaces. And in chemistry, reactions that are very rare in solution can readily occur when the reactants become concentrated.

But what about the actual polymerization reaction? In PCR, the energy driving the reaction is provided by the chemical energy of the substrates, which are the four nucleoside triphosphates: ATP, TTP, GTP, and CTP. As each of these in turn binds to the active site of the polymerase, a phosphodiester bond is formed to link it to the growing chain of nucleic acid, and a pyrophosphate molecule is released. This brings us to the question of a prebiotic source of energy to drive polymerization reactions by condensation of monomers with loss of water molecules.

REPLICATION REQUIRES A SOURCE OF ENERGY

Synthesis of phosphodiester bonds that link nucleotides into polymeric nucleic acids depends on chemical energy. The usual approach to driving prebiotic polymerization reactions is to chemically activate the monomers so that spontaneous polymerization in solution becomes possible. This is what living cells do today. Amino acids are activated when they are attached to tRNA, and the chemical reaction that links amino acids through peptide bonds is energetically downhill. Nucleotides are activated by synthesizing them as NTPs. The polymerization reaction catalyzed by a polymerase enzyme is also energetically downhill, with the release of pyrophosphate as the leaving group. But there is a problem with the simulations that use RNA as a template. NTPs don't work very well in the non-enzymatic polymerizations described above.

Now we come to a major gap in our understanding of the first replicating molecular systems: No one has yet found a plausible way to chemically activate nucleotide monomers so that they can undergo polymerization on a template. One of the reasons for writing this book is to suggest how to bridge this gap—so that a set of molecules can begin to replicate on early Earth in the absence of enzyme catalysts and activated nucleotides. *If there is no obvious way to activate nucleotides by adding a high-energy bond, let's try to discover a plausible condition in which water itself becomes a leaving group.* This idea is not new, but, in fact, was proposed in the 1970s by David Usher at Cornell University, who suggested that dry heat could serve as a source of energy to drive phosphodiester bond formation. Usher and his student Angelika McHale went on to demonstrate that such a bond could be produced between two hexamers if they were lined up on a template having complementary base pairing. McHale and Usher's observation has not been followed up by others, but the use of dry heat as a condensing agent is likely to be fruitful, as I will describe in Chapter 14.

THE FIRST REPLICATING SYSTEMS

By this time, you know that I enjoy Greek and Latin derivations of scientific and technical terms. I tell my students that knowing a derivation helps us to remember what words mean and to use them with precision. Words also undergo a kind of evolution that is analogous to biological evolution. For instance, a neologism, from the Greek words meaning "new word" is a symbiotic linking of two older words. Most neologisms that are proposed don't find a niche and become extinct, while a few, such as "software" and perhaps "googling" become firmly embedded in our language. Speciation of languages occurs when cultures become separated by distance. The so-called romance languages of French, Italian, and Spanish all started as one language perhaps 5,000 years ago and are now separate species. There are even mutations that we begin to see in a century or two. Examples in British and American spelling include *centre* and *center* and *whilst* and *while*.

There are often surprises when we discover a derivation, some of them actually fun. For instance, the word "pancreas" comes from Greek words meaning "all meat." (There are no bones in the pancreas.) And the word "gastrocnemius," referring to the bulging calf muscle, means "belly of the leg." Then there is *Drosophila melanogaster*, the fruit fly attracted to rotting bananas, whose name means "black-bellied garbage lover."

But what about the word "replicate"? There are three Greek words that permeate the life sciences. One is *spiro*, meaning "breath," from which we get inspire, expire, perspire, transpire, and respire—all having to do with breathing. (And don't forget spirit, the breath of life.) The second is *plica*, having the sense of folding. This gives us replicate (to fold repeatedly) and duplicate (folding once so that one becomes two) from which we get duplex DNA, the double helix. Other related words are complicate (life is complicated), implicate, and explicate. Finally, there is the Latin word *evolvere*, meaning "to unroll." To revolve is to roll repeatedly, to involve is to roll in, and to evolve is to roll out. So there you have it. The early scientists who invented these words or applied them to scientific matters had a working knowledge of Greek and Latin, and they used it as precisely as possible.

The winning entry in the origins of life sweepstakes seems to be RNA, closely related to DNA. The reason is that RNA not only can transmit genetic information and does so in all life today, but it can also act as a catalyst. Chapter 11 told this story and described some of the remarkable properties of ribozymes, or RNA catalysts. One of RNA's most significant properties is that it can undergo a kind of evolution in the test tube, as demonstrated by Jack Szostak and Jerry Joyce in the 1990s. But can we say that RNA is the final answer, the molecular solution to life's beginnings? We have come closer, but not quite to the finish line. The problem is that even RNA seems too complex to have been the first polymer involved in life processes. Furthermore, we have not found a way by which RNA can evolve on its own. Instead, it needs the help of protein catalysts. But most scientists in the field would place their bets on something resembling RNA as the first true molecule able to reproduce itself. And when RNA got together with simple proteins to produce primitive ribosomes, the system was well on its way to becoming the first living organism.

Now we come to the main question of this chapter: How could replication have started up on prebiotic Earth in the complete absence of any catalytic enzymes? We don't know yet, but there are some significant clues, one of which was discovered by Leslie Orgel working at the Salk Institute in La Jolla, California. The Salk, as it is affectionately known, was founded in 1960 by Jonas Salk, who developed the first vaccine to the polio virus and became an American scientist-hero. I recall as I was growing up in Santa Monica how much my mother feared that her three boys might get polio, which attacks the spinal nerves and can produce lifelong paralysis. Several of my colleagues from that era did come down with polio and still get around in wheel chairs. In severe cases, breathing is affected, and children had to be kept alive in "iron lungs" that did their breathing for them. I also recall getting my first polio shot as a child, greatly relieving my mother's anxiety. Salk's vaccine, and a second oral vaccine developed by Albert Sabin, virtually extinguished polio as a health threat in the United States.

When Leslie Orgel joined the Salk faculty in 1964, he had already established his research on the origins of life, and he maintained this interest until his death in 2007. In the late 1970s, Orgel and his student Rolf Lohrmann thought that dry heat might drive nucleotide polymerization, and they published a paper showing that this condition could in fact produce dimers and trimers. However, this was not a very elegant process if you are a biochemist. There must be a way to make polymerization work in solution. Orgel decided to try attaching a chemical leaving group called imidazole onto the phosphate, and this proved to be a key to a wonderful advancement in the study of replicating systems. When they are involved in chemical reactions, leaving groups live up to their name. Imagine a phosphate and a ribose brought close together by binding to a surface. It is possible for an ester bond to form to produce ribose phosphate, and in this case, water itself is a leaving group. However, this reaction is energetically uphill, and only a few ester bonds are produced before the reaction comes to equilibrium. But now imagine that imidazole replaces one of the hydroxyl groups (–OH) on the phosphate. The imidazole bond to the phosphorus has higher chemical energy than the bond to the hydroxyl group, so it is more likely to come off to leave an exposed phosphorus ready to react with something. If this occurs with a ribose nearby, one of the hydroxyl groups on the ribose immediately forms a chemical bond with the exposed phosphorus in what organic chemists call a nucleophilic attack. The result is a spontaneous, energetically downhill formation of the ester bonds required to make a nucleic acid polymer. In a very real sense, life depends on metabolic reactions that install leaving groups on monomers to activate them. Phosphate and pyrophosphate most often serve as leaving groups in the biosynthesis of polymers.

In a series of papers with his colleagues Rolf Lohrmann, Alan Schwartz, and Gerald Joyce, Orgel demonstrated template-directed non-enzymatic polymerization of RNA. The first experiments used polycytidylic acid as a template and an imidazole ester of guanosine monophosphate (GMP) as the monomer. These were chosen because G forms Watson-Crick base pairs with C, but also because G, a purine, tends to help the reaction along by what is called stacking energy. The large, flat purine bases composed of two rings tend to associate into stacks at higher concentration, while smaller pyrimidine bases with a single ring do not. The stabilizing effects of hydrogen-bonded base pairing and stacking forces meant that the activated guanosine monophosphate would line up along a strand of polycytidylic acid. Because the imidazole is a potential leaving group, when the activated phosphate nudged up against a neighboring phosphate, the imidazole fell off and a phosphoester bond could form.

In a typical reaction, the solution was left to incubate for a week at 0°C. The low temperature helped the guanosine to maintain hydrogen bonds on the template, and it also reduced the rate at which the activated nucleotide spontaneously hydrolyzed in solution. When the products were analyzed by HPLC, a beautiful set of peaks could be seen, indicating that the GMP had polymerized on the template, producing RNA products ranging up to 30 or more nucleotides in length. In the absence of the template, virtually no long-chain products were observed. Furthermore, the reaction was very specific. If activated nucleotides other than GMP were used, they did not get incorporated into polymers.

With this method in hand, a series of other experiments became possible. For instance, it was soon discovered that certain metal ions like lead and zinc acted as catalysts to further improve the yields. For a while, everyone thought that we might finally understand how replicating systems of molecules could arise in the prebiotic environment. No enzyme was required to catalyze polymerization. All that was needed was a template of RNA and some way to activate monomers that could use the template to guide their polymerization into a second strand. However, what I described above is not a true replication, but instead represents the synthesis of a second strand that is a complementary copy of the template strand (not a duplicate of it). For instance, the system works well if polycytidylic acid (polyC) is the template so that polyguanylic acid (polyG) is synthesized. To duplicate the original polyC, the polyG should also be able to act as a template for its synthesis, but this could never be made to work in a satisfactory way. The reason is that pyrimidine bases in nucleotides like cytidine monophosphate (CMP) and uridine monophosphate (UMP) are single ring structures, while the purine bases of guanosine monophosphate GMP and adenosine monophosphate (AMP) have two rings. The pyrimidine rings do not produce stable stacks, and are therefore unable to stay in one place long enough on the polyG template for the polymerization reaction to occur. So far, this has remained an intractable problem and awaits another breakthrough discovery if we are to understand how replicating molecules could begin to function in a primitive version of life.

CONNECTIONS

Replication of genetic information in DNA was invented by early forms of life, and most researchers would consider replication to be a primary feature by which we define life. An important point is that errors can occur during replication. The altered base sequences then produce the variations in a population of organisms that are essential for evolution by natural selection, which is the other unique property of life.

However, it is equally clear that the complex process by which DNA is replicated today could not have been the original version of replication, so a simpler, more primitive mechanism remains to be discovered. One possibility is that RNA was the first replicating molecule because it can act as a catalyst in the form of ribozymes but also carries genetic information in the sequence of bases. It seems possible that someone will fabricate an RNA molecule with a nucleotide sequence that allows it to fold into a ribozyme capable of catalyzing its own reproduction in a test tube environment. This will be an extremely important discovery, front page news, in fact. But keep in mind that the ribozyme could not reproduce unless it were part of a system that included a container, a continuous supply of monomers, and a process by which those monomers are chemically activated so that they have sufficient energy to undergo polymerization. In the laboratory, the container is a test tube, and the monomers are already activated by the investigator.

But there were no test tubes or investigators on prebiotic Earth. Is there a polymer even simpler than RNA that had the ability to replicate and evolve under prebiotic conditions? This is where our current knowledge ends, so the search is on. Although at first it might seem impossible to discover a process that emerged four billion years ago, in principle it may be relatively simple. A set of monomers must have physical properties that allow them to organize themselves into linear arrays by self-assembly or perhaps be organized on a surface. The monomers must also have specific chemical properties so that a source of energy can cause water molecules to leave and produce ester bonds that link the monomers together in chains. Finally, the chains must be able to act as templates that bind monomers in a specific sequence in order to ensure the next cycle of replication. Although at some point, catalysts evolved to speed replication, the first replicating polymers were probably able to carry out the process by themselves.

My guess is that the system of replicating templates and monomers, whatever its composition, resembled the polymerase chain reaction described in this chapter. The PCR reaction requires cycles of hotter and cooler temperatures, and these would be available as fluctuating conditions in the prebiotic environment. However, PCR also requires chemically activated monomers. Was there a plausible prebiotic source? Maybe, but we haven't found one yet. Perhaps there is an easier way to produce polymers which does not require activated monomers, but instead uses fluctuating conditions to provide energy for ester bond formation. One possibility will be described in Chapter 14.

13

HOW EVOLUTION BEGINS

Nothing in biology makes sense except in the light of evolution.

THEODOSIUS DOBZHANSKY, 1973

The ability to evolve is the final aspect of any definition of life. Whatever system of molecules we imagine to be the first form of life or fabricate in the laboratory as synthetic life must have a demonstrated capacity for evolution. As a graduate student working on biophysical problems having to do with membranes, I had no particular interest in questions related to evolution, but in 1967 I joined the Zoology faculty at the University of California, Davis, where Francisco Ayala was an assistant professor in genetics. Francisco was a former student of Theodosius Dobzhansky when he was at Columbia University, and invited Dobzhansky to serve as a visiting professor in the Department of Genetics at UC Davis, where he spent his remaining active years. I would occasionally chat with Dobzhansky, but I was young and naïve, just beginning my career, and had no idea that this friendly, elderly man was one of the great evolutionary geneticists of the twentieth century. Now, 40 years later, this chapter takes its theme from the title of one of Dobzhansky's last publications: "Nothing in Biology Makes Sense Except in the Light of Evolution."

How in the world could something like an evolving living system appear out of nowhere on a sterile early Earth? This is so mind boggling that a few scientists (very few) simply give up and state flatly that it had to be done by an intelligent designer and that we can never understand how it came about. I'm more optimistic, of course. We don't know all the answers yet, but we are making undeniable progress.

Does the origin of life make sense in the light of evolution? It does because life's beginning was also the beginning of biological evolution, and even before life began, there were simpler versions of evolution involving chemical and physical processes that occurred spontaneously in the prebiotic environment. Although Charles Darwin knew virtually nothing about molecular biology, he did consider that evolution began with the origin of life. In Darwin's great book, *The Origin of Species,* he asks the question we are addressing in this chapter:

> Looking at the first dawn of life, when all organic beings, as we may believe, presented the simplest structure, how, it has been asked, could the first steps in the advancement or differentiation of parts have arisen?

Darwin's book provided the first clear statements of the fundamental principles of evolution. Let's put ourselves in Darwin's shoes, and use his insight to consider how evolution could begin. First of all, we're talking about forms of life much simpler than even the most primitive bacteria today. Darwin would tell us that evolution could not begin with a single organism and that we need to find a way to generate large numbers of primitive organisms in the prebiotic environment. Furthermore, there must be considerable variation in their properties. The requirement for variation within a population means that the first life forms capable of evolution were not simply a mixture of reproducing molecules, but instead consisted of microscopic systems of interacting molecules encapsulated in some sort of boundary structure—what we now call cells. Next is that the cells had to find a way to grow. The way life does this now is to take in simple molecules called nutrients from the environment and then use energy to link them into the polymers we call proteins and nucleic acids. Last, the cells needed to store genetic information and replicate it when they reproduced so that their properties were passed along to the next generation, but not perfectly. In other words, there needed to be a constant injection of a certain amount of errors, what we now call mutations, so that life could explore different niches and begin the long trek to today's biosphere.

THE ORIGIN OF LIFE CAN BE UNDERSTOOD AS AN EVOLUTIONARY PROCESS

A massive weight of evidence supports evolution as a fact. In what follows, I will briefly sketch the major milestones of evolutionary theory and then describe how we can demonstrate evolutionary processes at the molecular level in the laboratory. This demonstration lets us better understand how similar processes could lead to systems of molecules on early Earth that were stepping stones to the first forms of cellular life.

Charles Darwin had no knowledge of nucleic acids as carriers of genetic information, nor did they understand how changes over time are caused by mutations in genes. What they did know is that patterns appeared when one observed variations in populations of living organisms, and the patterns indicated that one species of organism, such as Darwin's finches in the Galápagos Islands, if geographically isolated, would begin to adapt to varying conditions until the original founding species produced multiple other species. Darwin also knew that over hundreds of years, breeding produced varieties of domestic animals and plants that were very different from wild species. A breeder simply selected a desired trait and then bred only those plants or animals exhibiting those traits. Several hundred to a thousand years of breeding could produce plants like tulips, roses, chrysanthemums, wheat, and corn, and domestic breeds of animals like sheep, cattle, and pigs. Pets in particular were heavily selected, an extreme example being tiny Chihuahuas and huge Great Danes—dogs so different that if they were discovered living in the wild they would at first be considered to be different species. Darwin concluded that species were produced by a kind of natural selection that begins when populations are separated in space and begin to adapt to different conditions.

Darwin also knew that fossils could be found in sedimentary rocks like limestone and sandstone. No one at the time realized how old the rocks actually were, but it was logical that older rocks would be deeper in the limestone, with younger rocks in layers nearer the surface. When fossils from older and younger rocks were compared, there was a clear trend from simpler organisms like corals and trilobites that dominated older rocks, to larger, more complex organisms with skeletons in younger rocks.

PHYLOGENETIC TREES

All of this led inexorably to the idea that evolution of life on Earth can be understood as a kind of tree. In his notebook from 1837, Darwin drew a very rudimentary tree showing how life could evolve from a primitive progenitor over time. A more detailed tree illustrating speciation was published in his 1859 book, *The Origin of Life*, and in my copy of the sixth edition, now over a century old, the tree folds out between pages 140 and 141. During the 100 years that passed following publication of Darwin's book, a number of other trees were proposed that were primarily based on obvious differences between organisms in terms of their morphology, nutrition, and behavior. In the biology textbooks that I learned from in the 1960s, the consensus was that life on Earth could be divided into five distinct kingdoms, and some biologists are still comfortable with this classification. Starting with the simplest, and using the scientific names, these were monera, protista, fungi, plantae, and animalia. In this system, the monera were bacteria lacking nuclei, and the protista were single-celled organisms with nuclei (protozoa). There was also a general understanding that living cells could be divided into prokaryotic bacteria that lacked nuclei, and eukaryotic cells that had nuclei.

Even though five kingdoms made sense, there were some odd discrepancies in the scheme. One of these had to do with chloroplasts and mitochondria. Chloroplasts are the light-harvesting organelles of green plants, but there are also photosynthetic bacteria that use chlorophyll to capture light energy. Mitochondria in both plant and animal cells synthesize ATP using electron transport as an energy source, but so do aerobic bacteria. In the 1960s, Lynn Margulis at Boston University proposed that all this would make sense if mitochondria and chloroplasts originated as endosymbiotic microorganisms. In other words, they were once free living, but at some point around two billion years ago, they merged with another single-celled microorganism and set up housekeeping, to the mutual benefit of both. Later, the genetic material of the first cell became organized into a nucleus, and the first eukaryotic cells emerged from populations that were entirely prokaryotic.

Margulis's idea was at first strongly resisted by traditional biologists, but then it was discovered that mitochondria and chloroplasts both had circular DNA and ribosomes that were clearly remnants of an original bacterial genome and protein synthesis apparatus. The nucleotide sequences in the DNA have now been traced back to cyanobacteria in the case of chloroplasts, and proteobacteria are the closest extant relative of the original mitochondrial species.

The mystery of where mitochondria and chloroplasts came from was solved, and the answer provided a much richer idea of how life evolved not necessarily as a perfect tree stemming from a single origin of life, but instead as a more complex tree in which some of the branches grow together and merge into new forms of life.

THREE DOMAINS

The next major advance in our understanding of the tree of life was put forward by Carl Woese at the University of Illinois, who realized that the RNA content of ribosomes might hold clues to the relationships between different microorganisms. When Woese compared the nucleotide sequences in the ribosomal RNA of different bacteria, it became clear that there was not just one general kind of bacteria, but two distinct forms of microbial life, so different that it would be necessary to invent a new classification system. In this system, life on Earth can be understood as three domains: the prokaryotic archaea and the eubacteria that lacked nuclei, and the eukaryotes that had nuclei. All three domains are rooted in an organism referred to as LUCA, or the last universal common ancestor. The LUCA organism does not necessarily represent the origin of life, but instead the microorganisms that emerged as the most successful either in competing for nutrients and energy or perhaps the most able to survive an extreme environmental stress that destroyed all competing forms of life. For instance, Norman Sleep at Stanford University suggested that life could have arisen multiple times in a variety of forms, but it then was subjected to the Late Heavy Bombardment described in Chapter 3. The vast energies released by impacting comets and asteroid-sized bodies wiped out all varieties of life in shallow water, but those microorganisms that happened to inhabit hot marine habitats resembling today's hydrothermal vents were able to survive. As a result, their particular genomes, which used a specific genetic code, emerged to become the universal genetic code.

Assuming that the LUCA concept was correct, Woese and his colleagues proposed a different kind of branching structure that encompassed all life on Earth. Figure 28 shows a tree with a root (the LUCA) in the center that separates into the three domains. The

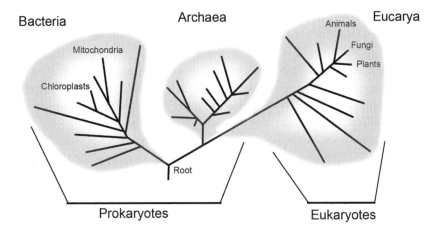

Bacteria   Archaea   Eucarya

Animals

Mitochondria

Fungi

Plants

Chloroplasts

Root

Prokaryotes   Eukaryotes

FIGURE 28

The three-domain concept, a single root branching into Bacteria, Archaea, and Eucarya. The diversity and biomass of life on Earth consists largely of microorganisms. Animals, including human beings, are represented by a small branch within the Eucarya, not far removed from plants and fungi in this scheme deduced from changes in ribosomal RNA sequences over time.

length of the lines indicates the number of changes per site on ribosomal RNA that separates each species from all the others, and the branches are deduced from the patterns of nucleotide sequences in the RNA.

The three-domain model, with its weight of evidence from ribosomal RNA, replaced the five-kingdom model that relied on differences in the physical characteristics of organisms. In the years following publication of the three-domain concept, advances in sequencing DNA became much easier, resulting in the publication of the entire human genome in 2000. Scientists then began to sequence not just animal and plant genomes, but also those of hundreds of microorganisms. As their results poured into the literature, it became clear that something was happening that might make it impossible to find an actual root for the tree of life. That something was a process by which bacteria could exchange genetic information. It had been known for years that bacteria were able to give off and take up small bits of circular DNA called plasmids, and this knowledge gave rise to an entire industry of recombinant DNA technology in which a specific gene could be inserted into bacteria as a plasmid and then expressed as a protein. But now DNA sequencing showed that it was not just plasmids. Bacterial genomes also contained genetic information that could only come from what is now called horizontal gene transfer (HGT). In other words, genes were not necessarily transmitted in a linear fashion between generations; they also were traded sideways between different miroorganisms. This discovery inspired Ford Doolittle, a Canadian researcher at Dalhousie University, to propose a new vision in which evolution began not as a tree growing out of a single original seed, but instead as an anastomosing network in which genetic information was more or less freely distributed.

Carl Woese agrees. Here is a quotation from the abstract of a paper he published in 2002:

> Aboriginal cell designs are taken to be simple and loosely organized enough that all cellular componentry can be altered and/or displaced through HGT, making HGT the principal driving force in early cellular evolution. Primitive cells did not carry a stable organismal genealogical trace. Primitive cellular evolution is basically communal. The high level of novelty required to evolve cell designs is a product of communal invention, of the universal HGT field, not intralineage variation. It is the community as a whole, the ecosystem, which evolves.

With the vast amounts of DNA sequence information now available, new versions of evolutionary trees are being proposed almost every year. They all incorporate the three-domain concept, but there is now an attempt to take HGT into account to the extent possible. More recent trees are based on a computational analysis of thousands of species, ranging from bacteria to humans. There is still a root in the center, the LUCA, with three branches, but now each of the branches proliferates into increased numbers of organisms.

The complete tree of all life on Earth can never be visualized because it would require a tree with millions of different organisms. The detailed organization of future trees will be so complex that it will probably exist primarily in addressable computational databases, rather than images, and will be undergoing constant additions and revisions as we gain new knowledge about the genomes of all life on Earth.

EVOLUTION IN THE LABORATORY

We can now address a simple question that is central to our understanding of the origin of life: Can non-living molecular systems evolve? Can genetic information really appear out of nowhere, by chance? If the answer is no, then we're in trouble because those of us who work on the origins of life claim that is exactly what happened four billion years ago, when the first forms of life emerged from a sterile mixture of minerals, atmospheric gases, and dilute solutions of organic carbon compounds. To answer that question, I will describe a classic experiment by David Bartel and Jack Szostak, published in 1993. Their goal was to see if a completely random system of molecules could undergo selection in such a way that defined species of molecules emerged with specific properties. Bartel and Szostak, working at Harvard Medical School, began by synthesizing many trillions of different RNA molecules about 300 nucleotides long, but the nucleotides were all present as random sequences. Making random sequences is easy to do with modern methods of molecular biology. Bartel and Szostak reasoned that buried in those trillions were a few catalytic RNA molecules called ribozymes that happened to weakly catalyze a ligation reaction in which one strand of RNA is linked to a second strand. The RNA strands to be ligated were attached to small beads on a column and then were exposed to the trillions of random sequences simply by flushing them through the column. This process could fish out any RNA molecules that happened to have even a weak ability to catalyze the reaction. Bartel and Szostak then amplified those molecules by an enzyme-catalyzed process and put them back in for a second round, repeating the process for 10 rounds. This is the same basic logic that breeders use when they choose to select a property such as coat color in dogs.

The results were astonishing (Figure 29). After only four rounds of selection and amplification, Bartel and Szostak began to see an increase in catalytic activity, and after 10 rounds, the rate was seven million times faster than the uncatalyzed rate. It was even

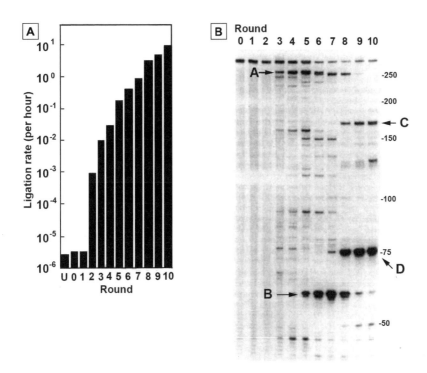

FIGURE 29

This figure, adapted from Bartel and Szostak's 1993 paper, demonstrates that genetic information can be introduced into a mixture of random RNA sequences by selection, amplification, and molecular evolution. The panel on the left shows catalytic ribozyme activity on a logarithmic scale. At the end of ten rounds, the activity has increased by a factor of 7 million. The panel on the right shows how molecular species appear and sometimes disappear in the mixture. Each of the dark bands is produced by RNA molecules that are being selected for catalytic activity. Two bands (A and B, indicated by arrows) appear in the third and fourth round but then "go extinct." At the end of the tenth round, two new bands 75 nucleotides and about 170 nucleotides in length (C and D, indicated by arrows) dominate the mixture of ribozymes. Modified from Bartel and Szostak, "Isolation of new ribozymes from a large pool of random sequences," *Science* 261 (1993): 1411–1418. Reprinted with permission from AAAS.

possible to watch the RNA evolve. Nucleic acids can be separated and visualized by a technique called gel electrophoresis. Here you see such a gel. At the start of the reaction, nothing could be seen, but with each cycle new bands appeared. Some came to dominate the reaction, while others went extinct.

Bartel and Szostak's results demonstrate fundamental principles of evolution at the molecular level. At the start of the experiment, every molecule of RNA was different from all the rest. There were no species, just a mixture of trillions of different molecules. But when a selective hurdle was imposed in the form of a ligation reaction that allowed only certain molecules to survive and be reproduced enzymatically, after a few generations, groups of molecules began to appear that displayed ever-increasing catalytic function. In other words, in a mixture that initially contained completely random RNA molecules, species of molecules appeared in an evolutionary process closely reflecting the natural selection that Darwin outlined for populations of living organisms. These RNA molecules were defined by the sequence of bases in their structures, which caused them to fold into specific conformations that had catalytic properties. The sequences are analogous to genes because the information they contained was passed between generations during the amplification process.

CALCULATING THE ODDS THAT LIFE COULD BEGIN BY CHANCE

It has often been proposed that life could not have begun by an accidental assembly of monomers into a polymer that happened to have the catalytic properties of an enzyme. The calculation for a ribozyme might go something like this: Assume that a particular ribozyme is 300 nucleotides long and that at each position there could be any of four nucleotides present. The chances of that ribozyme assembling are then 4^{300}, a number so large that it could not possibly happen by chance even once in 13 billion years, the age of the universe. The Bartel and Szostak experiment directly refutes this conclusion, which rests on the assumption that selection and amplification do not occur. The inescapable conclusion is that genetic information can appear out of random mixtures, as long as there are populations containing large numbers of polymeric molecules with variable sequences of monomers and a way to select and amplify a specific property.

A similar process must have occurred on prebiotic Earth to bring the first forms of life into existence. The milestones I presented here and in earlier chapters suggest that both compartments and polymers could be produced on early Earth and that the cellular compartments would have no trouble capturing samples of the polymers to produce a near infinite variety of protocells. In my lab, we make cellular compartments by adding water to a few milligrams of dry lipids in a flask. A milky suspension is produced that contains trillions of individual microscopic vesicles in the size range of small bacteria, 0.5 micrometer in diameter. If the vesicles are prepared in a solution containing small peptides and short nucleic acids such as RNA, each of the trillion vesicles will contain a different set of components.

Now let's think about early Earth. Instead of milligrams of lipids in a flask, early Earth would have had billions of tons of organic material assembling into enormous numbers of microscopic compartments, and half a billion years of time to do the experiment. The origin of life is best understood as a metaphor of combinatorial chemistry, but at a level far beyond what we can do in the laboratory. Will we ever discover the combination of ingredients that gave rise to life? Again, I am optimistic. We need to apply what we know about the chemistry and physics of living systems to reduce the odds and then be brave enough to actually try the experiments. In the next chapter, I will propose one such experiment that has not yet been performed, but which I think will be very revealing about life's beginnings.

In science, it is rare to have ideas that are all your own. Usually someone else has had similar ideas, sometimes in publications 150 years old, but other times from colleagues whose thought pattern happens to parallel your own. Here is what Charles Darwin wrote that first captured the idea of evolution proceeding by a process of selection:

> But if variations useful to any organic being ever do occur, assuredly individuals thus characterised will have the best chance of being preserved in the struggle for life; and from the strong principle of inheritance, these will tend to produce offspring similarly characterised. This principle of preservation, or the survival of the fittest, I have called Natural Selection. It leads to the improvement of each creature in relation to its organic and inorganic conditions of life; and consequently in most cases, to what must be regarded as an advance in organization. Nevertheless, low and simple forms will long endure if well fitted for their simple conditions of life.

CONNECTIONS

Now I can summarize the series of linked concepts that I have described both here and in previous chapters, which will set the stage for the hypothesis to be proposed in the next chapter. The prebiotic environment had a variety of energy sources and simple carbon compounds that could drive chemical reactions toward greater complexity. These reactions produced ever more complex carbon compounds, some of which could assemble into membranous compartments, while others could link up into polymeric chains. The polymers became encapsulated in the compartments to produce vast numbers of protocells. The variable protocellular compartments are a microscopic version of what we now call combinatorial chemistry. Each protocell contained a different mix of polymers and monomers, and represented a natural experiment. Some of the polymers happened to have potential catalytic abilities, while others could in some way replicate. We haven't reproduced this process in the laboratory yet, but several research groups are working on it. The main point is that a rare few of the cellular compartments contained both catalysts and replicating molecules in which the catalysts could speed up the replication, and the replicating polymers could carry a kind of genetic information that coded for the monomer sequence in the catalysts. It is conceivable that both properties were present in a single molecular species resembling RNA.

Biological evolution can begin when there are compartmented systems of molecules that can grow and reproduce. At first, there would be no competition, and the only selective factor would be which particular system can make the most efficient use of the energy and nutrients available in the environment. However, at some point, the resources would become limiting, and a variety of environmental stresses would begin to affect the primitive populations of microorganisms. Examples of stresses include variations in temperature, pH, ionic content, mechanical disturbances such as turbulence, and ultraviolet light that can produce photochemical damage to nucleic acids. The competition for resources and ability to cope with environmental stress would then become selective factors. Within the populations of primitive microorganisms, there would be variations in their abilities to compete for resources and to withstand stress. One of the most effective strategies would be to find ways to cooperate so that resources and protective measures are shared within populations. It follows that microorganisms would quickly evolve mechanisms to produce a second layer of complexity in the form of systems called microbial mats in which populations of bacteria cooperate for mutual benefit, rather than compete.

14

A GRAND SIMULATION OF PREBIOTIC EARTH

I suggest that the minimum requirement for an organism capable of stepwise
evolution therefore consists essentially of a group of paired semi-periodic solids which
determine the order of polymerization in another group of unpaired semiperiodic
solids of different backbone structure which, by their catalytic and carrier activities,
make available the required concentrations of the monomeric precursors of the
genetic and catalytic polymers . . . I suggest that the occurrence of proteins and
nucleic acids as the main types of semi-periodic solids found in all living organisms
that have been examined may be governed by the fact that other types of materially
and energetically inexpensive semiperiodic substances do not display the
complementariness required to give efficient replication.

PETER MITCHELL, 1957

I was a high school student in Ohio when Peter Mitchell wrote the carefully worded sentences quoted above. They are taken from a brief note he published in the proceedings of the first major international conference on the origins of life, which convened in Moscow in 1957. Looking down the list of participants, it amazes me to realize who had made the arduous trip to this meeting long before there was convenient international travel on jet airplanes. The attendees included young Stanley Miller, who just a few years earlier had published his sole-author paper that showed how amino acids could be synthesized by sparking a simulated atmosphere of early Earth. Melvin Calvin also attended. Calvin was in the process of working out how carbon dioxide could be taken up by plants during photosynthesis, research for which he was awarded the Nobel Prize in 1961. Other past and future Nobelists were among the participants, including Harold Urey (1934), Wendall Stanley (1946), Linus Pauling (1954), Ilya Prigogine (1977), and Peter Mitchell (1978). John Desmond Bernal, Norman Pirie, Erwin Chargaff, Hans Fraenkel-Conrat, and Alfred Mirsky were also there, all of whom had made significant contributions to our understanding of how life works at the molecular level.

Aleksandr Ivanovich Oparin himself presided at the meeting. Oparin sparked research on the origins of life with his pioneering 1924 book, and the conference was in part organized to honor his contribution. But among all of these superb scientists, Mitchell alone formulated the clearest vision of life's beginning, and his prescient idea can be taken as the central concept underlying what I will say in this chapter.

All scientific progress builds on the foundations set in place by previous workers. Isaac Newton is often quoted as noting that "if I have seen further it is by standing on the shoulders of giants." Peter Mitchell was one of the giants in my life, but the ideas in this chapter rest on other shoulders as well, including those of Harold Morowitz at George Mason University and Victor Kunin, now at Lawrence Berkeley National Laboratory. Morowitz wrote a book called *Beginnings of Cellular Life: Metabolism Recapitulates Biogenesis*, in which he clearly stated why the first life must have been enclosed by membranous compartments we call cells. Furthermore, he showed how energy flowing through protocells could drive the evolution of primitive metabolic pathways. Kunin published a paper in 2000 entitled "A System of Two Polymerases," in which he proposed that the first life must have incorporated a pair of interacting polymerase enzymes, one composed of protein, the other of RNA. Edward Trifonov and T. Bettecken in 1997 had suggested a similar idea, but Kunin's paper enlarged the concept first outlined by Peter Mitchell in the quotation that begins this chapter. I also owe a debt to Noam Lahav, who in the 1980s while working at NASA Ames, introduced me to the idea that fluctuating conditions could drive condensation reactions required for polymer synthesis. In 1997, Noam and Shlomo Nir outlined the basic concepts, and Noam published his own book in 1999 with the title *Biogenesis: Theories of Life's Origins*.

My thinking was also influenced by David Usher at Cornell University. In the mid-1970s, Usher published two papers that greatly impressed me. His main point was that condensation reactions can occur if the reactants are exposed to anhydrous conditions at elevated temperatures. Usher and his student Angelika McHale published a paper in 1977 that demonstrated synthesis of a phosphodiester bond by this mechanism. They added two short oligomers of RNA—six nucleotides each—to a solution containing a complementary template 12 nucleotides long. A phosphodiester bond was formed between the two neighboring ends when the dry molecules were exposed to a temperature equivalent to desert heat, which added activation energy to the reactants.

This chapter incorporates these earlier ideas, but I will also propose a way to test them. The test is related to three unanswered, yet fundamental questions regarding the last stage before life begins:

- How could polymers like nucleic acids and proteins be synthesized in the absence of metabolism and enzymes?
- How could the polymers replicate?
- How did the fundamental interaction between proteins and nucleic acids begin?

HYPOTHESIS, EXPERIMENT AND THEORY

We can now go through the logical process of fabricating a conceptual framework and then show how it can be tested. The basic steps in the hypothesis involve self-assembly of membrane compartments that add order to otherwise disordered reactants, synthesis of polymeric molecules, encapsulation of the polymers in the compartments, uptake of nutrients and energy, and estimating the chances that the polymers will happen to have catalytic properties and the ability to store and transmit genetic information. The first step is an exercise called framing the hypothesis, in which we list the assumptions that provide its foundation and boundaries. Here are the assumptions outlined as a set of bullet points:

- All life today is cellular. Although subsystems such as ribozymes and polymerases display certain functions of life, no subsystem has all the features required to fit a definition of life. For this reason, an experimental test of the hypothesis will incorporate the potential for cellular compartments to form.

- The primary molecules of life are polymers, and the growth that is characteristic of life occurs when monomers are assembled into polymers by an enzyme-catalyzed energy-dependent process. There were no enzymes before life began. Therefore, we must discover mechanisms by which polymers can be produced non-enzymatically from plausible monomers existing on prebiotic Earth.

- All cellular life depends on cyclic processes in which two species of polymers interact with each other. Polymeric catalysts called enzymes guide the synthesis of nucleic acids, and the base sequences in nucleic acids guide the sequence of amino acids in the enzymes. This means that life began not simply as a mixture of polymers in a bag, but as a system of molecules that happened to self-assemble in the prebiotic environment.

- RNA by itself is not sufficiently complex to do all the things we ask of first life. A new approach is required in which amino acid polymers and nucleic acids co-evolve within compartments. The main question to be addressed is how polymeric catalysts and information-carrying molecules could arise spontaneously and begin to interact so that selection and evolution become possible.

- Life began when countless numbers of encapsulated molecular systems were produced by random synthesis of polymers followed by self-assembly in membranous compartments. Evolution began when the few such systems were able to grow and compete for resources, while mixing genetic information in such a way that useful genes were combined into new systems of interacting polymers.

The overall aim of the hypothesis is to involve not just one, but two kinds of interacting polymers. The way to test the hypothesis is to fabricate a prebiotic simulation that can run experiments under conditions that fulfill the following essential requirements: The components are sufficiently complex so that two kinds of polymers are continuously produced from mixtures of monomers; multiple cycles allow complexity to increase step-wise in the system; and the system incorporates a version of combinatorial chemistry so that evolutionary selection can occur at the molecular level.

PREBIOTIC SIMULATIONS REVISITED

There is a tremendous emphasis on hypothesis testing in my kind of science. When I serve on panels to review grant proposals for NASA, NSF, and NIH, one of the first things we look for in a proposal is a clearly stated hypothesis, followed by critical experimental tests. This emphasis is peculiar to science and even defines the primary method of science. I don't know of any other human activity where hypothesis testing plays such a major role in judging the value of a proposed action. The only thing that comes close, perhaps, is the kind of thinking that goes into buying stocks. A smart investor looks at the properties of a given company and then compares that record to other companies, looking for patterns of profit and loss. The investor then makes a prediction about the future value of the company (a hypothesis) and can test the prediction by buying stock (an experiment). By analogy, a scientist sees a pattern in the natural world and comes up with a hypothesis that might explain that pattern. The hypothesis is then tested by experiment, and the investment is the time and energy spent by the researcher. Once in a while, we even get a return on our investment.

The word "hypothesis" comes from Greek words meaning "below theory," and that is exactly right. A hypothesis is not a theory but an idea that might be right or wrong. Only when a hypothesis has been tested experimentally, and only when the results are considered to have general significance, is it elevated to the status of a theory. Some theories, particularly in the physical sciences, incorporate what we call laws. A law is a statement of an observation that has never failed, such as the laws of thermodynamics that I described in Chapter 6. This can even penetrate popular culture, as in the bumper sticker that reads "Gravity: Not just a good idea. It's the law!" Most of us have heard of the major theories that have emerged from the scientific enterprise and affect our lives in multiple ways. Newton's theory of gravity, Darwin's theory of evolution, and Einstein's theory of relativity are classic examples. Newer theories include quantum mechanics, the Big Bang, and stellar nucleosynthesis of elements.

THE SCOPE OF ALTERNATIVE HYPOTHESES

What about the origins of life? There is no theory yet, but there are dozens of hypotheses. What I will describe as a test of the hypothesis is a simulation of early Earth at the time of life's origin. The fact is that virtually all the progress in understanding life's beginnings relies on simulations. The classic example is Stanley Miller's simulation discussed in Chapter 4, in which a mixture of gases including hydrogen, methane, ammonia, and water vapor was exposed to an electric spark. The result revolutionized our thinking because surprising amounts of biologically relevant compounds such as amino acids were unexpectedly synthesized under these conditions. Other examples include the synthesis of adenine from hydrogen cyanide, discovered by John Oro, and the synthesis of simple carbohydrates from formaldehyde in alkaline solutions. Then there are physical simulations that involve interactions between mineral surfaces that catalyze certain synthetic reactions, such as clay surfaces, iron-nickel sulfide in boiling water, and the high-pressure, high-temperature reactions that occur when pyruvic acid is heated in gold capsules. (See Table 6 for a summary of simulations.) A common feature of all these simulations is that they are designed to be simple in terms of their starting conditions. This permits the investigator to study just one kind of reaction and to limit potentially confusing complications from other reactions that might occur. Researchers are trained to keep things simple, so they instinctively shy away from more complex mixtures. Furthermore, research depends on funding from public granting agencies. Because a design for a realistic prebiotic simulation is expensive, and the outcome uncertain, a complex simulator has never been constructed and tested.

Earlier simulations often focused on producing the monomers of life, but what about polymers? Life did not begin with simple mixtures of pure compounds reacting in distilled water at room temperature. Instead, life emerged from a highly complex mixture of monomers in aqueous solutions that interacted with mineral surfaces while being subjected to a variety of energy fluxes. The simulator I will sketch out below incorporates features of the prebiotic environment that have been discussed earlier in this book and forces us to think about complexity in terms of plausibility arguments. I think that it is time for such a simulator to be constructed because we will learn a great deal by watching the evolution of interacting monomers and polymers under conditions that are sufficiently complex simulations of the prebiotic environment.

So, what would such a "grand simulation" look like? The actual container would not need to be very large, perhaps 1 cubic foot of volume. It needs this much space just to accommodate the components of the simulation, which include mineral surfaces, controlled atmospheric gases, fluctuating conditions of wetting and drying, different compositions of added organic solutes, and varying energy sources. It is also a small enough volume so that it can be sterilized before the experiment begins. The mixture of organic compounds in the chamber is a nutrient medium, and if a single live bacterium got in, months of work could be lost. One side of the cube will have a silica glass window that will permit light to be added both for illumination and as a potential energy source in some of the experiments. The container needs to be sealed so that the atmospheric gas composition can be controlled, but it must also have access ports for adding components, taking samples, and flushing gas through the chamber. The entire volume needs to be temperature-controlled from 0°C to 90°C. A dedicated computer and software will run the experiments, which will require hours to days or even weeks to carry out. The computer will control temperature, gas flow, and composition, and it will inject water, take samples, and monitor parameters such as temperature, pH, and relative humidity.

I estimate that the simulator could be constructed in a year, at a cost of $500,000, half of which is salary support for an instrument design engineer, a control circuit engineer, and a machinist. The next step, analysis of the organic compounds present, gets much more expensive. The reason is that the analysis will follow the chemical changes that occur over time, and this requires several costly instruments, including high-performance liquid chromatography (HPLC), mass spectrometry, capillary electrophoresis, and electron microscopy. A dedicated laboratory incorporating these instruments would cost in the range of several million dollars, and would require the salaries

TABLE 6 Examples of Previous Prebiotic Simulations

Publication	Reactants	Products	Energy Source	Conditions	Simulation
Miller, 1953	CH_4, H_2, NH_3	Amino acids	Electric discharge	Gas phase reaction	Atmosphere
Oro, 1961	HCN	Adenine	Chemical	High pH	Solution chemistry
Fox & Harada, 1958	Amino acids	Polymers	Dry heat	180°C	Geothermal
Ponnamparuma, 1972	HCHO	Sugars	Chemical	High pH	Solution chemistry
Huber & Wächtershäuser, 1997	CO, CH_3SH	C–C bonds	Chemical	100°C	Mineral catalysts
McCollom et al., 1999	CO, H_2	Hydrocarbons, fatty acids	Chemical	200°C	Geothermal
Matsuno, 1999	Glycine	Oligoglycine	High temperature	Hydrothermal vent, quench	Hydrothermal vents
Ferris et al., 2002	Nucleotides	RNA	Chemical	Clay catalyst	Mineral catalysts
Orgel, 1983	Nucleotides	RNA	Chemical	Solution chemistry	Template-directed
Cody, Hazen, et al., 2002	Pyruvate	Complex mixture	Chemical	High pressure, high temperature	Geothermal
Usher & McHale, 1976	RNA hexamers	RNA dodecamers	Anhydrous conditions	Desertlike temperatures	Template-directed
Allamandola et al., 1990s	Water, methanol, ammonia	Organic product mixture	UV light	High vacuum, low temperature	Dust grains, interstellar medium

of two full-time technicians to run the samples and maintain the instruments, which would amount to another million dollars over a five-year period that the experiment would run.

Of course, this is only a thought experiment, which costs nothing, but you can begin to understand why a simulator has not yet been constructed. We refer to this kind of project as high-risk, high-payoff research. It's high risk because nothing may come of it, but it is high payoff because if we get lucky, we will answer one of the great remaining questions of biology.

SIMULATING THE PREBIOTIC ENVIRONMENT

Now let's think about what should be put into the cube, and how we would run an actual experiment. There are at least three levels of complexity, which are outlined in Table 7, and all three would need to be run to understand what is happening in the simulation. The first level is essentially a control experiment because no organic compounds will be initially present. The question is: What happens when a simple mixture of gases is exposed to an energy source in the chamber? We would not expect much to happen, other than the synthesis of small organic compounds in trace amounts.

A very important second experiment would also be run at this level, which is to add known polymers of nucleic acids and peptides. It is essential to determine the rate at which phosphodiester and peptide bonds are hydrolyzed under the conditions of the experiment. The reason is that in later experiments, we expect to see such polymers accumulate in the chamber, and if the rate of hydrolysis is such that they break up in a few minutes, very little net synthesis will be observed. It will be essential to adjust experimental parameters, such as temperature and pH, so that hydrolysis rates are measured in hours, days, or even longer.

The second level of complexity is to assume that on early Earth, various highly reactive compounds were synthesized and could react. In the simulation, we would add small amounts of compounds, such as hydrogen cyanide and formaldehyde, and again run the experiment, monitoring what happens over time.

Quite a few experiments similar to the first and second levels of complexity have already been performed by earlier researchers, but the third level of complexity is the next step, the actual experiment in which it is assumed that the monomers of biopolymers were present as a mixture in the prebiotic environment. Later I will discuss possible outcomes, but first we need to go into more detail about the components and physical conditions being simulated. The discussion to follow focuses on Level 3 complexity.

A WATERY ORIGIN

Liquid water was essential for life to begin, but what kind of water? Was it acidic? Basic? Salty or fresh water? The most plausible guess, in my judgment, is that life began in

TABLE 7 Levels of Complexity in the Simulation

Level	Atmosphere	Aqueous Phase	Mineral Phase	Energy Sources	Expected Products or Possible Outcome
1. *Example of simplest conditions*	Carbon dioxide, nitrogen	Fresh water (pH~5) or sea water with ferrous chloride and phosphate	Volcanic lava	UV light, electric discharge, anhydrous cycles	Expected products: trace amounts of organic compounds
2. *Examples of reactive compounds added to Level 1*	Carbon dioxide, nitrogen, hydrogen sulfide, carbonyl sulfide	Fresh water or sea water, HCN, HCHO, phosphate	Volcanic lava	UV light, electric discharge, anhydrous cycles	Expected products: amino acids, traces of purines, sugars, monocarboxylic acids
3. *Examples of monomers added to Level 2*	Carbon dioxide, nitrogen	Fresh water or sea water, six amino acids, four mononucleotides, ribose, phosphate, nucleobases, fatty acids, fatty alcohols, glycerol	Volcanic lava, clay	Wet-dry cycles (fluctuating environment, temperature 60°C–900°C)	Possible outcome: synthesis of polymers, self-assembly of protocells that encapsulate interacting systems of polymers, potential for simple catalysts to arise

moderately acidic fresh water. It was acidic for two reasons. First, the atmosphere had a significant content of carbon dioxide compared to the present atmosphere, and whenever carbon dioxide dissolves in water, it produces a weak acid called carbonic acid. Even today, if a glass of pure water is left open, the pH will be neutral at first (pH 7), but after a few days, it will become more acidic as the very tiny amount of atmospheric carbon dioxide dissolves in the water, finally reaching pH 5.5. The second reason is that volcanic gas contains sulfur in its gaseous forms of sulfur dioxide (SO_2) and hydrogen sulfide (H_2S), which also form weak acids when they dissolve in water.

The question of salt water or fresh water is more difficult, and my plausibility argument is more controversial. The consensus today is that life began in a marine environment, simply because most of Earth's water is salty sea water. On the other hand, when my colleagues do experiments in their laboratories related to the origin of life, they almost never use sea water. The reason is that the high salt content of sea water often interferes with the reactions they are investigating. Instead, they use very pure water produced by special ion exchange devices, and buffers are added to keep the pH in a neutral range. Such controlled conditions are intended to minimize the variables that otherwise make experiments hard to interpret.

Would there be sources of fresh water on early Earth? It seems reasonable to assume that at some point, volcanic land masses would rise above sea level and begin to collect precipitation. The rain would interact with active volcanism to produce the kinds of hot springs we see today in Hawaii, Iceland, and Kamchatka.

To sum up, the simulation would initially use fresh water, not sea water, and it would be enclosed so that the atmosphere can be a mixture of carbon dioxide and nitrogen. Naturally, in testing the simulation, a good research plan does not make just a single assumption, but instead tests the assumption by varying the conditions. In an actual experiment, we would first work with fresh water, but then increase the content of sodium chloride, calcium, magnesium, and ferrous iron in increments to see what effect the variables have on the outcome. In this way, we would learn much more about the constraints of the system. My expectation is that higher concentrations of such ions will increasingly inhibit reactions that are important for the origin of life, but I could be wrong, so it is necessary to test my expectation.

MINERAL INTERFACES

What minerals should be present in our simulation? There are several possibilities, all of which were described in previous chapters. Bob Hazen thinks that chiral surfaces such as calcite may tip the balance toward the chirality that living organisms now use: L-amino acids and D-sugars. Gunther Wächtershäuser has proposed that the reaction of iron and sulfide during pyrite mineral formation could provide a source of electrochemical energy to drive primitive metabolic reactions. Jim Ferris has presented clear evidence that certain clay surfaces are able to organize nucleotides in such a way as to enhance their polymerization into short segments of RNA. All of these proposals involve chemical properties and reactions on solid mineral surfaces, but an alternative idea was proposed by Michael Russell, a geologist at Cal Tech, who suggested that the porous structures of minerals composing hydrothermal vent formations have microscopic compartments that provide some of the advantages of cell boundaries without invoking the actual assembly of lipid protocells.

It is not yet certain whether mineral interfaces are essential for the origin of life, but it does seem reasonable to choose a plausible mineral to interact with the organic components. Otherwise, by default, we will have chosen glass as the mineral substrate, which would not be a plausible prebiotic mineral surface. The most abundant minerals that would have been present on early Earth would simply be the minerals associated with volcanic activity, so I would choose to introduce sterile samples of lava with depressions in their surfaces, where small puddles of water can accumulate and evaporate. The reason for the puddles will be described later.

ADDING ORGANIC COMPOUNDS

The next step in developing the hypothesis is to decide what organic compounds should be present. At first, this seems not too difficult to work out. There are only four basic kinds of molecules in living organisms: carbohydrates, fatty compounds, proteins, and nucleic acids. Why not just add them all? This would be a bit silly because those four compounds are all highly evolved substances produced by metabolic processes in life today. We might learn something useful, such as how fast they break down under simulated prebiotic conditions, but we would not get much information about the origin of life, which is a process that goes in the other direction, from simple mixtures to complex interacting systems.

We could try something a little bit smarter, which is to use the organic material present in carbonaceous meteorites to simulate the kinds of molecules that would have been available on prebiotic Earth. We know that the Murchison meteorite contains more than 70 different kinds of amino acids, a few carbohydrates, some lipidlike compounds, and even trace amounts of nucleic acid bases, such as adenine. Small amounts of phosphate are also present, mostly attached to organic molecules. This mix contains everything required for life, and my New Zealand colleague Michael Mautner has shown that bacteria and plants can grow in a solution containing nothing but organic material from the Murchison meteorite as a source of nutrients.

However, carbonaceous meteorites are rare, and are extremely valuable for research purposes. It would take many kilograms of the Murchison meteorite to get enough organic material to do our experiment, but there are only a few kilograms available for research. This means that we can be guided by our knowledge of meteoritic organics, but we must use our wits to choose a subset of representative compounds for the simulator.

For these reasons, only the most plausible amino acid species will be present in the basic experiment. The majority of the 20 amino acids present in proteins are synthesized by specialized enzymatic metabolic processes, but six amino acids are relatively abundant both in meteorites and in the products of Miller-Urey syntheses. These are glycine, alanine, valine, aspartic acid, glutamic acid, and proline. Should we add racemic mixtures of the amino acids or just the homochiral forms that all life uses today? I think at first we should choose L-amino acids, saving the question of the origin of chirality for later studies.

What else might we add? All life is based on nucleic acids, so their components should be present, which would be adenine, guanine, cytosine, and uracil, the bases of RNA. We would add ribose and phosphate as well because these are the linking groups that hold the bases together in long strands. Again, we have a choice to make. No one has yet worked out a way to make nucleotides starting from bases, ribose, and phosphate, so we should probably add the nucleotides, assuming that there was some way for them to be synthesized on early Earth. This, by the way, is just a hopeful guess. There might have been a much simpler way to make replicating molecules that we have not yet discovered, but again, we must start somewhere, and nucleotides are the only game in town.

Finally, and easiest, is the choice of lipids to make membranes. We will add a mixture of fatty acids with chain lengths from 12 to 14 carbons long, along with glycerol and phosphate, so that simple membrane-forming lipids can be synthesized by the condensation reaction that John Oro, Will Hargreaves, and I reported in 1977.

To sum up, Table 7 shows a plausible mix of organic substances for the simulator at complexity Level 3. Of course, this is just the bare bones level of complexity, and there is no guarantee that it will be complete. Guided by the initial results, it is likely that additional components will be introduced in order to establish degrees of sufficient complexity for the system.

ENERGY SOURCES

Suppose that we have constructed our simulator and are ready to begin. We have a steaming puddle of hot water bathing porous lava, and all the organic compounds of Level 3 complexity are present. What will happen?

The answer is, not much. Maybe there will be a few chemical changes taking place, such as cytosine undergoing deamination to become uracil, but for the most part when we analyze the mixture a day, a week, or a month later, it will be the same mix of compounds that we started with. Why is that? The reason has to do with thermodynamics. Everything in the system is at equilibrium, and the only way to drive it away from equilibrium is to add an energy source. What kind of energy should be added?

Here is where we need to do some careful thinking. The problem is that every chemical reaction is reversible, so if we add enough energy to make something interesting happen, whatever it is that happens can also go in the other direction, back toward equilibrium, which is much less interesting. This fundamental fact is characteristic of all life. To illustrate, we can consider Louis Pasteur's experiment in which he ground up some meat in water, boiled it, and filtered the solution into a closed flask. This clear soup has all the nutrients required for life, perhaps equivalent to 1 gram of amino acids and other small soluble biochemicals extracted from the meat, but it is sterile from the boiling and filtration. However, if we add a single live bacterial cell, it immediately begins to use the nutrients and energy to grow and divide. Half an hour later, there are two bacterial cells, then four, then eight, and two days later, there are so many bacteria that the originally clear soup has become cloudy. If we could weigh all the bacteria, there would be about 1 gram of living matter. The result is a vast increase in complexity, from a single bacterial cell to trillions of cells, each containing several thousand different kinds of proteins required for life.

Inevitably, it can't last. The bacteria use up all of the available chemical energy and nutrients, so they stop growing and dividing. Over the next few months, the individual bacterial cells die and begin to disintegrate, and the solution again becomes sterile. If we could wait long enough (many years), the polymers composing the bacteria would undergo hydrolysis and return to equilibrium, but most of the energy originally stored in the nutrients is gone. Where did it go? The carbon compounds have been metabolically degraded to waste products (such as carbon dioxide, which contains virtually no chemical energy) because their energy content was released by metabolism and lost to the outside world as heat. The point of this thought experiment is that life requires energy to drive it toward complexity, and a living system such as a bacterial culture can only be maintained in a steady state away from equilibrium if there is a continuous source of energy.

Now we can return to the original question of what energy source should be added to the simulator. Only five energy sources would have been available on early Earth: heat, light, chemical energy, electrical energy, and radioactivity. Of these, we can exclude ordinary electrical energy and radioactivity immediately. Both have far too much energy content to be useful in driving the chemical reactions leading to life, even though electrical energy in the form of lightning may have helped synthesize basic compounds like amino acids. For the same reason, we can exclude light in the ultraviolet range and heat energy at temperatures much above the boiling point of water.

That leaves ordinary light, heat energy, and chemical energy. Let's look at these more closely. The biosphere runs almost entirely on chemical energy produced by plants, which capture light energy from the sun. The process of photosynthesis uses light energy to pull electrons out of abundant water and then donates those electrons to carbon dioxide to form high-energy sugars and fats with oxygen as a by-product. We, and other denizens on the animal side of the equation, take in the chemical energy stored in nutrients, and our metabolic pathways allow the electrons to run back downhill to oxygen, thus releasing the energy for growth, movement, nervous activity, and all other energy-dependent processes.

So, should we add light energy to our simulator? My guess is that light is probably not a primary energy source driving the origin of life. The reason is that plants use a pigment—chlorophyll—in the initial capture of light energy, but there is nothing in our system so far that can serve as a pigment. We can't just add chlorophyll, because it is a fairly delicate molecule synthesized by complex enzyme-catalyzed reactions, and no one has yet proposed a plausible prebiotic version of a photosynthetic pigment system. How-ever, it would be interesting to add light energy at some point in our simulation just to see what happens. Light, particularly in the ultraviolet range, could drive photochemical reactions that might unexpectedly produce a pigment having primitive chlorophyll-like functions, and that would be a major discovery.

This leaves us with heat energy and chemical energy. At least for the initial produc-tion of the first complex protocells, chemical energy is not going to help much. For instance, we might add compounds containing chemical energy, such as ATP and the other nucleoside triphosphates used by cells to make nucleic acids. In the absence of enzymes that catalyze polymerization, these would not spontaneously form phosphodi-ester bonds. Even if we added a mineral such as clay, which is known to catalyze syn-thesis of RNA from activated nucleotides, everything would run back downhill toward equilibrium when the reaction was over. That means that we would need to constantly add energized compounds to the system to balance the energy being used up.

The primary source of energy is one of the central enigmas of origins of life research. Several simulations of prebiotic conditions can produce polymers from activated mono-mers, but it is not a simple matter to come up with a plausible constant source of energy-containing molecules. The one exception is the reaction in which carbonyl sulfide (COS) activates amino acids to form peptide bonds, but so far this has been shown only to link just a few amino acids rather than producing long peptide chains.

We have now eliminated everything except heat energy, and in fact, cycles of anhy-drous heating would be my choice for driving the uphill reactions leading to protocells. It is not an original idea. Sidney Fox and his colleagues were among the first to investi-gate this energy source, and they discovered that polymers could be produced simply by heating a dry mixture of amino acids. In his early work, Leslie Orgel also succeeded in making small yields of short RNA molecules in this way.

Now I need to recall a point made in Chapter 6, which is that heat is *not* a primary energy source. No life on Earth today directly captures heat energy in the same way that light and chemical energy are used. However, heat has two important properties that are related to energy use. Let's consider a common reaction, such as drying a sugar solu-tion to produce caramel candy. If we allow a sugar solution to dry at room temperature, it forms a clear, hard, glassy solid—or even crystals—but only if the drying process is slow enough. Add a little yellow dye and lemon flavor, and you have lemon drops. Now let's dry the sugar by boiling the solution on a stove at 100°C. Instead of a clear solid we end up with a brown sticky material called caramel, which is composed not of pure sugar, but instead of a polymer in which the sucrose molecules have joined together. Heat is not driving the reaction, which is spontaneous due to the energy content of the sugar molecules. Instead, heat adds activation energy, which makes it more probable that neighboring sugar molecules will overcome an energy barrier and polymerize.

The second way that heat can be involved in primitive chemistry has to do with con-centration effects. Everyone knows that if we boil water, it evaporates to dryness, and if there is anything in the water, it becomes highly concentrated during the drying process. The most important point is that when something becomes dry, all of the free water molecules have been driven off into the surrounding atmosphere. Depending on the kinds of compounds present in the water, there is another form of water still available, which is the water that can potentially be produced by chemical reactions between the concentrated molecules. To illustrate such a reaction, suppose we put together a mixture of ethanol and acetic acid in a one-to-one ratio and then let it sit for several days to reach equilibrium. At the end of that time, if we analyzed the solution, we would find that a fraction of the two compounds had reacted according to the equation below, and that some ethyl acetate, an ester, has appeared along with some water. Furthermore, if we heated the solution, the reaction would go much faster and reach equilibrium in a few minutes. This kind of reaction, in chemical terminology, is called condensation:

$$\text{ethanol} + \text{acetic acid} \{\Rightarrow\} \text{ethyl acetate} + H_2O$$

Now we come to the central point of our hypothesis, which was described in detail in Chapter 4. *Most of the chemical bonds linking monomers into the polymers of life are produced by the removal of water from the monomers.* Examples of chemical bonds resulting from condensation reactions include the ester bonds of fats and phospholipids, the glycoside bonds linking carbohydrates into polymers like starch and cellulose, the phosphodiester bonds of nucleic acids, and the peptide bonds of proteins. In the polymerization reactions of all life today, water is removed from potential monomers by enzyme-catalyzed metabolic reactions, and only then are the molecules able to react. We refer to this process as activation of the monomers. The central energy supply driving such reactions originates in the removal of water from adenosine diphosphate and phosphate to produce adenosine triphosphate, or ATP, the so-called energy currency of life. This reaction is carried out by chloroplasts using light energy, and by bacteria and mitochondria using the energy of electron transport from metabolites like acetic acid to oxygen.

$$ADP + PO_4 \{\Rightarrow\} ATP + H_2O$$

The chemical energy stored in ATP is then used to remove water from amino acids when they are linked to transfer RNA, and only then can ribosomes attach the amino acids to a growing peptide chain during protein synthesis. ATP also transfers its chemical energy to other nucleotides such as GTP, CTP, TTP, and UTP, and polymerase enzymes use them as substrates to synthesize nucleic acids, the other fundamental polymer of living systems.

A central problem for understanding the origins of life is that no one has discovered anything like ATP that might serve as an energy currency to drive other reactions. Herrick Baltscheffsky at Stockholm University has proposed that pyrophosphate might serve this purpose, and Belgian Nobelist Christian DeDuve suggests that thioesters also could act as a source of chemical energy, but so far it is not clear how pyrophosphate or thioesters could be produced in the prebiotic environment and used to drive polymerization reactions. What we need is an abundant source of free energy that is continuously available and highly plausible and that can produce the polymers required for the kinds of natural experiments that would lead to the origin of life. This brings us back to making polymers by drying, concentrating, and heating because we know with certainty that it is possible to make polymers this way.

Is it that simple? Can we produce life just by the "shake-and-bake" method? Most of my colleagues don't think so, and for very good reasons. In general, if we dry and heat a mixture of monomers, all sorts of chemical bonds form, and the result is a brown or black goo called tar. Heat-driven reactions are too messy because products are so diverse. Another concern is that all living organisms run their reactions in an aqueous medium, so chemists interested in the origins of life prefer their reactions to occur in solution.

On the other hand, there is no doubt that the energy provided by heating and drying can drive the synthesis of long, potentially interesting polymers. What this means is that if we can find a way to organize monomers so that specific bonding is promoted and then have them undergo condensation reactions under anhydrous conditions using heat for activation energy, we could produce virtually all the polymers involved in life processes. For this reason, our simulation will incorporate cycles of wetting and drying expected to occur wherever land, air, and water formed an interface on early Earth.

This is not a new idea, and it has often been explored by earlier researchers. We call this a fluctuating environment, and it is a way to "pump" a chemical system uphill toward greater complexity. Polymers are formed by condensation during the dry portion of the cycle and then dispersed and mixed during the wet phase. The new idea to be described here is that the reactants and products will not be free in solution, but instead will be encapsulated in microscopic compartments produced by the amphiphilic molecules that are present in the mixture.

ENCAPSULATION PROCESSES

Now I need to return to the idea of combinatorial chemistry. When a biotechnology company wants to explore a reaction that produces some useful product, it no longer relies on lonely chemists laboring far into the night doing one experiment after another in an attempt to optimize the reaction. Instead, it employs a kind of robotic device that delivers small amounts of reactants into thousands of tiny compartments on a plastic plate. Each compartment differs from all the others in parameters such as pH, ionic strength, ratios of reactants, and so on. The reaction is allowed to take place, and the products are analyzed to see which particular set of conditions works best. By doing thousands of experiments in parallel, the robot can do in a week what used to take years of effort by chemists performing experiments in series, one experiment per day.

I think that a natural version of combinatorial chemistry was involved in the origin of life. There were no robots on early Earth, so how can a chaotic mix of energy sources and organic compounds spontaneously generate the conditions of combinatorial chemistry? The answer lies in the fact that the unit of all life today is the cell, and it is easy to make cellular compartments under simulated prebiotic conditions. As described in Chapters 7 and 8, given a source of amphiphiles, the presence of membranous compartments seems inevitable. Furthermore, under the drying-wetting conditions described above, it is also a simple matter to produce not just compartments, but to fill them with whatever compounds are present as solutes. During the drying cycle, the dilute mixtures are highly concentrated into very thin films within the amphiphilic mass in which chemical reactions can proceed. Not only would the compounds react with one another under these conditions, but the products of the reactions would become encapsulated during the wet phase of the cycle. The result of this process would be vast numbers of what we call protocells that appeared all over early Earth, wherever water solutions were undergoing wet-dry cycles in volcanic environments similar to today's Hawaii or Iceland. The protocells are compartmented systems of molecules, each different in composition from the next, and each a kind of natural experiment required for combinatorial chemistry.

This all sounds rather simple, but the payoff of a successful demonstration would be immense. Why has no one yet attempted to build a simulator? The reason is that most scientists prefer relatively simple conditions with just one or a few components so that they can control and study a specific reaction. This is the accepted style of doing research, and the idea of putting together an incredibly complex simulator with multiple sources of energy and large numbers of components just to see what happens will not appeal to most peer review committees. They would refer to the project as a fishing trip, a phrase that is sure to kill a grant proposal. The expectation would be that we would be lost in the immense complexity of all possible reactions.

Well, maybe. But I am inspired by the boldness of young Stanley Miller when he added energy in the form of electrical discharge to a mixture of reduced gases. His advisor, Harold Urey, was very skeptical, predicting that Miller would only produce "Beilstein," referring to a famous compendium in the 1950s that listed all known organic compounds, which numbered in the thousands. Instead, something else happened. In a perfect example of emergent phenomena, substantial amounts of amino acids were produced, and the result sparked a revolution in our thinking about the origins of life. (To be sure, thousands of minor products were also synthesized, and even today these have not been thoroughly analyzed.) Miller's experiment is an example of what I refer to as the principle of sufficient complexity. If he had left out a single component in his system, for instance ammonia, it would have been insufficiently complex, and nothing very interesting would have happened.

Summing up, I think we should be bold and go fishing in conditions that we judge are sufficiently complex, with the expectation that whatever happens in the simulation, we will learn from it. We may even discover something about life's origins beyond all expectations, simply by being willing to accept the origin of life as an emergent phenomenon arising from an incredibly complex environment and by taking the chance that similar reactions can be reproduced in the laboratory by trusting the laws of chemistry and physics.

Part of the enjoyment of doing research is that ideas pop into your head all the time. Everyone has ideas, but the hard part of research is to choose which ideas are worth spending time on and then performing critical tests that have the primary aim of destroying your beautiful idea. Why destroy? Because that way you can quickly discard bad ideas, and most of them are bad. Only after an idea survives the crucible of such testing can you begin to take it more seriously and submit it to even further testing. Then, if it survives, you can publish.

What I have described above is the history and rationale underlying an idea about how life can begin, and now it is time to test it. What I will next describe arises from my research on lipids. There are three aspects of the idea:

· Polymerization by condensation reactions is enhanced by ordering effects of lipid matrices.

· Lipid vesicles containing mixtures of polymers represent natural experiments in combinatorial chemistry.

· Encapsulated polymers can interact with one another—one as a catalyst and the other as a primitive gene.

It has been known for years that amphiphilic molecules such as fatty acids and phospholipids can form vesicles, as described in earlier chapters. It is also known that when the vesicles are dried, the resulting dry material is not just amorphous goo, but instead the vesicles fuse to produce an ordered structure. The two basic structures are called lamellar and hexagonal arrays. These are often referred to as liquid crystals because the molecules are not necessarily in a solid form, but instead can move about within the structure even though they are organized in such a way as to have crystalline properties. We also know that if a solute is present in the solution during drying, the molecules become trapped between the lamellae or cylinders of the organized lipid matrix, as I described in Chapter 8 on encapsulation.

Here is the new idea, which is central to the hypothesis being tested: *What if the lipid structures impose order on the monomers in such a way that specific polymerization is promoted during condensation reactions?* If so, such conditions would be expected to lead to the synthesis of long polymers. Furthermore, instead of adhering to a mineral surface, the polymers would be back in solution as soon as water is added to the system, and they would be encapsulated in lipid vesicles. Figure 30 shows how we imagine nucleotides would look when they are organized and concentrated between the lipid bilayers in a multilamellar lipid matrix. It is easy to see how order can be imposed on the nucleotides and why the ordering makes it much more likely that they will be able to form phosphodiester bonds that link them into nucleic acid molecules. When the nucleotides are in solution, they are in random motion, and each is surrounded by thousands of water

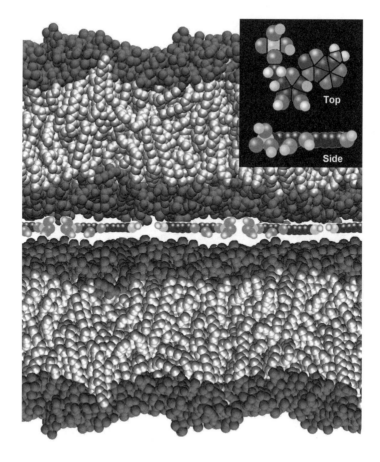

FIGURE 30

Adenosine monophosphate (AMP) is one of the nucleotide monomers of nucleic acid, and a molecular model of AMP is shown as top and side views in the inset. The main figure illustrates how AMP can become organized within the two-dimensional plane between lipid bilayers in a multilamellar lipid matrix.

molecules; but in the lipid matrix, the water is gone, and the molecules are confined within a two-dimensional plane.

Figure 30 illustrates the idea, but it is more than just an idea because we have some preliminary yet promising results that I will now describe. First, I want to say something about the way a research laboratory works in a university setting. Lab groups are very international, and we work as a team composed of graduate students, post-doctoral associates, technicians, and what is called a principal investigator, or PI. The PI's role is to generate the basic research ideas and approach and to convince a funding agency like NASA to fund the research with a grant, which is typically in the range of $100,000 per year. This is enough to support two or three people in the lab, most often a graduate student and a post-doctoral associate, a kind of apprentice who already has his or her PhD and is spending a couple of years learning how to be an independent investigator. For the work I will describe here, Dr. Sudha Rajamani (from India) did much of the hands-on work, Dr. Felix Olasagasti (from Spain) did some of the preliminary studies, Seico Benner (from Japan) and Amy Coombs (from California) carried out several of the technical assays, and Dr. Alexander Vlassov (from Russia) did some of the analytical gels that finally convinced us there was a real phenomenon. My job, of course, was to come up with the original idea (which I first began to work on 25 years ago) and to write the proposal that got the grant that supported this team for two years.

The basic experimental approach is fairly straightforward, but it had to be repeated quite a few times under different conditions to be convincing. We made up dispersions of lipid vesicles in water (actually a dilute acid at pH 3 to simulate the acidity of volcanic hot springs). Mononucleotides, the monomers of RNA, were added so that there was about one nucleotide per lipid in the solution. The mixture was warmed to 90°C for 2 hours while it was dried with a gentle stream of carbon dioxide to simulate the prebiotic atmosphere. A small amount of water was then added, the mixture was stirred for a few seconds to disperse the lipid vesicle, and the cycle was repeated up to seven times. The idea was to simulate the conditions of a volcanic hydrothermal area on early Earth, similar to the sites in Kamchatka that I described earlier, in which a continuous drying-and-wetting process occurs at the edges of the pools. The water would be fairly hot (80°C to 90°C) and weakly acidic.

Well, what happened? We had a wonderful surprise. When we tested the solution for the presence of polymers, we found that RNA-like molecules had been synthesized, ranging from 20 to 100 nucleotides in length. The yields were very low by the standards of organic synthesis: less than 0.1% of the nucleotides had been linked up into longer polymers, representing a few micrograms of product from the milligram quantities of nucleotides present in the mixture. However, I see no particular reason to think that there were high-yield polymerization reactions occurring in the prebiotic environment. The reactants were complex mixtures of hundreds of different organic compounds, so the reactions leading to early biopolymers almost certainly had very low yields.

An important outcome of the experiment was that when the reaction was over and the last cycle of hydration was completed by adding water, the lipid captured the RNA in vesicles. I consider these vesicles to be the first step toward cellular life, that is, microscopic membrane-bounded compartments containing complex mixtures of polymers that have the potential to be both catalysts (ribozymes) and carriers of information.

But we can't get too excited yet. It's good to find a plausible way to make polymers resembling RNA, but they are not like biological RNA because the monomers are linked not only by 3'-5' phosphodiester bonds (the bond of biological RNA) but also unnatural 2'-5' bonds. This mix of bonds means that the molecules will not readily fold into structures that can have catalytic activity like a ribozyme. Furthermore, there is no obvious process by which the molecules can reproduce themselves. What we need to do now is to isolate and amplify those few strands that have catalytic activity. We are also running a second series of experiments, testing an obvious possibility that arises from the results described above. If it is possible to drive polymerization of nucleotide monomers into polymers like RNA by using lipid matrices to organize the monomers, the next step is to see whether the same conditions could drive template-directed synthesis. This follows the discovery by Leslie Orgel and his coworkers that a template can direct the synthesis of RNA in aqueous solutions as long as the monomers are chemically activated. We are now including synthetic nucleic acid templates in the mixture of mononucleotides and lipid vesicles. During drying, it is possible that the monomers will line up on the template to form phosphodiester bonds. If this works, it would represent a very plausible way for replication to begin on prebiotic Earth, driven non-enzymatically by anhydrous condensation reactions and directed by the sequence of bases in the template strand. Significantly, such a process would be the equivalent of a prebiotic polymerase chain reaction because the elevated temperature when water is added back to the dry matrix would cause the double-stranded nucleic acid to melt back into two separate chains, just as it does in the PCR devices we use every day in the laboratory. The result would be growth and replication of the original polymers, together with amplification of the sequences in the nucleic acids.

This, of course, would be a very exciting result, but there would still be more to do. The last step is to find conditions that permit the cycle of life to begin, in which a genetic molecule directs the synthesis of a catalytic molecule, and the catalyst, in turn, speeds up the synthesis of the genetic molecule. This is where our current knowledge ends and speculation begins. But speculation can turn into a useful hypothesis, and the simulation I described earlier in the chapter can provide the test.

BACK TO THE SIMULATION

From earlier research and the results described above, we know that anhydrous cycles can promote condensation reactions to produce polymers from nucleotides and amino acids. Furthermore, the resulting polymers exist in a state that is thermodynamically uphill because hydrolysis begins to break them down when they are exposed to water. However, the downhill reaction is slow in comparison to the uphill reaction, so the polymers will have a transient existence in a kinetic trap. Growth of polymers is just one aspect of what the origin of life is all about. There must also be a mechanism by which the polymers can reproduce and undergo selection for catalytic activity. Today, those polymers are universally proteins and nucleic acids. We don't yet know what the original polymers were, but we can be certain that at some point, two kinds of polymers started to interact in a mutually beneficial way. A reasonable place to begin experimental studies is to assume that one of the polymers was composed of amino acids and the other of nucleotides. Only then would the mixture of polymers captured in protocells be on the pathway to the origin of life.

This is where the second idea comes into play, first suggested by Peter Mitchell in the quotation that begins this chapter, and later elaborated by Victor Kunin as the two-polymerase scenario. We assume that the origin of life is an example of an emergent phenomenon, and by definition an emergent process depends on the chance interaction of components of a complex system. What we know about the origin of life is that the first living cells assembled somewhere on Earth within a time interval of half a billion years or less. What we don't know is whether the chances were so slim that it required an entire planet and half a billion years to occur just once. At the other extreme, if we happen to choose the right components and conditions of our simulation, the chances might be so good that we could see a form of primitive yet recognizable life arise in a few weeks or months.

We can now propose some experimental tests that require a complex simulation. We can assume nucleic acids will be produced by lipid-enhanced polymerization, and we can also assume that peptide bonds are formed under these conditions to produce short peptide chains having potential catalytic activity. If a mixture of mononucleotides and amino acids is present, both will undergo polymerization into longer polymeric structures. At some point, by chance, encapsulated polymers of short peptides and nucleic acids will be produced that happen to interact in such a way that synthesis of both kinds of polymers is promoted. When this happens, the vesicles containing those particular polymers will have an advantage over other vesicles lacking this interaction and will take over the "culture" of vesicles.

Now we come to the most important prediction to be tested. In the simulation at Level 3, after some number of cycles, we will begin to see not just a mix of many different RNAs and peptides, but instead certain peptides and RNA species will begin to dominate the mixture. This is what happened in Bartel and Szostak's experiment described in Chapter 13 on evolution. Even though trillions of random RNA molecules were present at the start of their experiment, after four cycles specific kinds of RNAs began to accumulate that had catalytic properties. I think that a similar process could occur in the Level 3 simulation, involving short strands of RNA around 50 nucleotides in length and short peptides containing about the same number of amino acids. No metabolism is required, no input of energy other than that provided by the wetting-drying cycles. The system is not yet alive because it still depends on the energy of wetting-and-drying cycles to drive polymerization and heat energy to activate the process. Each encapsulated vesicular system is different from all the others, so each represents a microscopic experiment. If enough experiments are done, at some point, one or a few of the vesicles will happen to have a grouping of polymers that can go to the next step, which is to capture an energy source other than dehydration energy of cycles.

Should we construct a grand simulation? I think we should because I doubt that we can ever understand the complex interactions of organic compounds and energy sources well enough to deduce a chemical pathway to the origin of life. Instead, we must have the naive courage that young Stanley Miller had and be willing to venture into the unknown by allowing encapsulated systems to emerge from sufficiently complex mixtures of plausible prebiotic compounds, thus discovering processes that are essentially unpredictable.

CONNECTIONS

Our understanding of self-assembly processes and conditions of early Earth's environment has advanced to the point that it is now possible to propose several possible pathways that lead from simple mixtures of organic compounds to the first molecular systems having the properties of the living state. In the scenario described in this chapter, membranous compartments appeared first on early Earth by self-assembly from amphiphilic molecules provided either by meteoritic-cometary infall or by geochemical synthesis in the high-temperature and high-pressure environments associated with volcanic activity. Not only were membrane compartments first to assemble, but they also were essential in assisting the synthesis of polymers resembling nucleic acids and peptides from available monomers. In this view, life did not begin as a specific molecule or solely as metabolism, but rather as a system of molecules held together in a membranous compartment that promotes polymerization processes.

A second feature concerns the original source of free energy to drive the molecular system toward ever greater complexity through polymerization. This involves the free energy made available by fluctuating environments, the most common being cycles of repeated wetting and drying that would naturally take place at the interface of bodies of water with land masses. A series of such cycles can "pump" polymerization processes by producing a chemical potential that favors loss of water from reactants so that ester bonds and peptide bonds are produced. This idea has a long history, but it was generally abandoned when it was discovered that the bonds were formed indiscriminately, leading to the production of intractable "tars." However, if a microenvironment could be found that limited the kinds of bonds that were formed, polymerization driven by cycles of anhydrous conditions would be a very plausible way to produce the initial systems of encapsulated polymers required as a first step toward cellular life. At some point, a particular system can then evolve an energy-capture mechanism that uses chemical energy or light energy. This will be a strong selective factor, so that particular system will rapidly overtake slower systems that depend on anhydrous cycles to generate polymers.

At this point, Darwinian selection would begin because each of the structures would have a unique set of specific components, and therefore would have varying abilities to undergo growth processes. The primary selective factor would simply be the rate and efficiency with which a given structure could capture energy and nutrient monomers. Successful versions would grow and at some point, become so large that a kind of division could occur. Each of the daughter structures would inherit some of the properties of the parent structure but would still be unique. More slowly growing structures would be left behind in the growth race, and structures incapable of growth would simply be a source of nutrients, slowly losing their components to growing structures.

15

PROSPECTS FOR SYNTHETIC LIFE

Invention, it must be humbly admitted, does not consist of creating out of void,
but out of chaos.

MARY WOLLSTONECRAFT SHELLEY, 1818

In Mary Shelley's classic tale, Dr. Victor Frankenstein assembled a human body from parts retrieved from corpses. The novel, first published nearly 200 years ago, raised questions that we would now consider to fall within the realm of bioethics. If Dr. Frankenstein wanted to carry out his experiment today, he would need to bring it to the attention of the IRB (Institutional Review Board) at his university who would doubtless reject it. And yet, a number of laboratories around the world are attempting to perform a reconstitution of life eerily similar to Frankenstein's dream—to invent something alive, but on a microscopic scale. There is even a name for such science: synthetic biology.

Here I will briefly trace the history of attempts to fabricate artificial cells that increasingly are approaching the definition of living organisms. These efforts have not yet succeeded, but there is reason to believe that the goal may be achieved in the next decade. The point is that as we attempt to assemble synthetic life, we are retracing some of the steps that led to the origin of life, and we can learn from successes and failures.

Assembling a system of molecules capable of reproduction was first achieved in 1955, when it was discovered by Heinz Fraenkel-Conrat and Robley Williams at the University of California Berkeley that the tobacco mosaic virus could be separated into its coat protein and RNA. Neither component was active by itself, but when mixed together, the two parts reassembled into the infectious agent. Although viruses are not considered to be alive in the usual sense of the word, a successful reconstitution of viral functions encouraged other investigators to attempt the reassembly of more complex systems. In the 1970s, Efraim Racker at Cornell University was the most successful practitioner of this art, using detergents such as deoxycholic acid to disperse membranous components of cells. When the detergent was removed, small membranous vesicles formed that contained the original components, albeit somewhat scrambled in their orientation. Despite the scrambling, Racker and his colleagues were able to reconstitute electron transport reactions and ATP synthesis of mitochondrial and chloroplast membranes.

Similar techniques were soon applied to other biological structures and functions. For instance, Walther Stoekenius and Dieter Oesterheldt at the University of California San Francisco reconstituted the proton pump of purple membranes isolated from a halophilic bacterial species that uses the energy of a proton gradient to synthesize ATP. In a remarkable collaboration, Racker and Stoeckenius teamed up to produce a membranous system containing both the proton pump of halobacteria and the ATP synthase of mitochondria, and they demonstrated that the hybrid membranous structures could synthesize ATP using light as an energy source. The paper that Racker and Stoeckenius wrote has often been cited as the publication that finally confirmed chemiosmotic synthesis of ATP, leading to a Nobel Prize for Peter Mitchell in 1978.

The point of this brief history is that relatively complex biological functions can be reconstituted by self-assembly of their dispersed components. Why not try to reconstitute a whole cell? If this turns out to be possible, perhaps it will help us untangle what we mean by "life" and even elucidate the major steps that led to the origin of cellular life nearly four billion years ago. Let's first consider what might happen if we disassembled a microscopic living organism and tried to put the pieces back together. We would not try this with something like amoebas because nucleated cells are too complicated. Instead, we should use a much simpler form of life, such as a tiny bacterium called *Mycoplasma*, a kind of microbial parasite that can only live in a relatively rich nutrient environment. Only 450 genes are present in its genome, while more complex bacteria such as *E. coli* have 10 times that number. The human genome, by way of contrast, is estimated to have 20,000 to 30,000 genes at last count, with the actual function known for less than half of these. The cells of *Mycoplasma* are bounded by a naked membrane composed of a mixture of lipids in the form of a bimolecular layer common to all membranes, with a variety of functional proteins and enzymes integrated into the bilayer. Some of the enzymes are responsible for extracting energy from nutrients and using the energy to synthesize ATP, while others are essential transporters of nutrients from the external medium into the cell. Examples of nutrients are an energy source such as glucose, amino acids as building blocks for proteins, and phosphate for nucleic acids. The interior contents of the cell include a circular strand of DNA having genes responsible for synthesizing proteins required for metabolism. A variety of structural components is also present, including thousands of ribosomes, the molecular machines that synthesize proteins, and hundreds of soluble enzymes involved in metabolism.

What happens if we add a detergent to *Mycoplasma*? The detergent immediately penetrates the lipid bilayer, which becomes unstable and breaks up into smaller particles containing lipids, membrane proteins, and the detergent molecules. With the membrane gone, the interior components are released. What we see visually is that the slightly turbid suspension of bacteria becomes clear, and if examined with a microscope, no cells are visible. The resulting solution contains all the components of the original living cell, but they have become diluted and disorganized.

Now we can try to reassemble the components by injecting a certain amount of order back into the system. This is done by a process called dialysis, in which smaller molecules like detergents leak through a porous membrane while larger molecules remain behind. (The same process is used to treat patients with kidney disease.) As the detergent leaves, the lipids self-assemble into bilayers that take the form of small vesicles. The membranous boundaries of the vesicles incorporate most of the functional enzymes and transport proteins that were present in the bacterial cell membranes, and each vesicle contains a random sample of the original contents of the bacterial cell, with one exception: The circle of DNA, the genome, that was originally packed tightly into the original living cells, has unraveled and is too large to be captured when the vesicles reassemble.

To be truly alive, the vesicles need most if not all of their original genes in the form of a DNA strand containing the genetic information required for 1,000 or so different proteins and RNA species, over half of which are the components of the ribosomes themselves. They would need genes for polymerase enzymes so that the DNA could be replicated as part of the growth process and a way for lipids to be synthesized because the membranous boundary must grow to accommodate the internal growth. Transport proteins must be synthesized and incorporated into the lipid bilayer, otherwise the vesicles have no access to external sources of nutrients and energy. A whole set of regulatory processes must be in place so that all of this growth is coordinated. Finally, when the vesicles grow to approximately twice their original size, there must be a way for them to divide into daughter cells that share the original genetic information.

THE SMALLEST CELLS

Well, how close are we to achieving true synthetic life in the laboratory? One way to get at this question is first to ask how small a living organism can be because in general, smaller means simpler, and it is best to first try making simple forms of life. In 1996, the images of the putative fossil bacteria in the Mars meteorite that I described in the Introduction were claimed to be in the range of 100 nanometers or less. This immediately raised suspicions among biologists, who knew that typical bacteria were approximately 10 times that diameter, which corresponds to a thousand-fold greater volume. Furthermore, ribosomes are 20 nanometers in diameter, which means that only a few could fit into such tiny cells.

This question spurred the National Academy of Science to convene a study committee with the specific charge of estimating the smallest version of life that is theoretically possible. The committee came up with an estimate of 250 genes in the DNA, and a few dozen ribosomes, all in a membranous compartment 200 nanometers in diameter. This is comfortably close to the smallest known form of life, *Mycoplasma*, with 450 genes.

Of course, this is life as we know it, which is based on a highly evolved relationship between DNA, RNA, ribosomes, and protein catalysts. It seems impossible that the first forms of life sprang into existence with such a complex system of interacting molecules, so perhaps smaller versions of exotic life are, in fact, possible. There are even controversial claims that something called nanobacteria exist everywhere and may cause certain diseases by producing deposits of a calcium phosphate mineral called apatite.

Evidence from phylogenetic analysis suggests that microorganisms resembling today's bacteria were the first form of cellular life. As described in Chapter 3, hints of their existence can be found in the fossil record of Australian rocks at least 3.5 billion years old. Over the intervening years between life's beginnings and now, evolution has produced bacteria which are more advanced than the first cellular life. The machinery of life has become so advanced that when researchers began subtracting genes in one of the simplest known bacterial species, they reached a limit of approximately 265 to 350 genes that appears to be the absolute minimum requirement for contemporary bacterial cells. Yet life did not spring into existence with a full complement of 300+ genes, ribosomes, membrane transport systems, metabolism and the DNA→RNA→protein information transfer that dominates all life today. There must have been something simpler, a kind of scaffold life that was left behind in the evolutionary rubble.

Can we reproduce that scaffold? One possible approach was suggested by the "RNA World" concept that arose from the discovery of ribozymes, which are RNA structures that had catalytic activities. The idea was greatly strengthened when it was discovered that the catalytic core of ribosomes is not composed of protein at the active site, but instead is only a tiny bit of RNA machinery. This is convincing evidence that RNA likely came first and then was overlaid by more complex and efficient protein machinery.

Another approach to the origin of life is now underway. Instead of subtracting genes from an existing organism, researchers are attempting to incorporate one or a few genes into tiny artificial vesicles to produce molecular systems that display all the properties of life. The properties of the system may then provide clues to the process by which life began in a natural setting of early Earth.

What would such a system do? We can answer this question by listing the steps that would be required for a living system to be reconstituted in the laboratory or to emerge as the first cellular life on early Earth:

1. Boundary membranes are formed by self-assembly of lipid molecules.

2. Macromolecules are encapsulated, yet smaller nutrient molecules can cross the membrane barrier.

3. The macromolecules grow by polymerizing the nutrient molecules.

4. The energy required to drive polymerization is contained in the nutrients themselves or supplied to the system by a metabolic process.

5. The energy is coupled to the synthesis of activated monomers which, in turn, are used to make polymers.

6. Some of the polymers are selected as macromolecular catalysts that can speed the growth process, and the macromolecular catalysts themselves are reproduced during growth.

7. Information is captured in the sequence of monomers in one set of polymers.

8. The information is used to direct the growth of catalytic polymers.

9. The membrane-bounded system of macromolecules can divide into smaller structures.

10. Genetic information is passed between generations by duplicating the sequences and sharing them between daughter cells.

11. Occasional mistakes (mutations) are made during replication or transmission of information so that evolution can occur by selection of variations within the population of cells.

Looking down this list, one is struck by the complexity of even the simplest form of life. This is why it has been so difficult to "define" life in the usual sense of a definition—that is, boiled down to a few sentences in a dictionary. Life is a complex system that cannot be captured in a few sentences, so perhaps a list of its observed properties is the best we can ever hope to do.

It is also striking that most of the individual steps have been reproduced in the laboratory. The task now is to integrate the functions into lipid vesicles to see how far we can get in reconstituting a living cell.

The process begins by encapsulating samples of cytoplasmic components from living cells like *E. coli*. As described in Chapter 7, encapsulation of large molecules is the simplest step toward the origin of cellular life. Pier Luigi Luisi and his coworkers at the Eigennössische Technische Höchschule in Zurich, Switzerland (usually and understandably referred to as ETH), made the first attempt to assemble a translation system by encapsulating ribosomes in lipid vesicles along with an RNA that codes for a specific amino acid, phenylalanine. The amino acid was attached to transfer RNA so that it could be used by the ribosomes. However, because the lipid bilayer was impermeable, ribosomal translation was limited to the small number of amino acids that were encapsulated within the vesicles. This is the kind of hurdle that is discovered when we attempt to assemble functioning cells. It forces us to consider mechanisms by which the first forms of cellular life could transport nutrients inward across their boundary membrane.

In 2004, Vincent Noireaux and Albert Libchaber, at Rockefeller University, published an elegant solution to the permeability problem. They broke open bacteria, again using *E. coli*, and captured samples of the bacterial cytoplasm in lipid vesicles. The aim of the experiment was to produce a functioning protein, so the complex mixture consisted of ribosomes, transfer RNAs, and the 100 or so other components required for

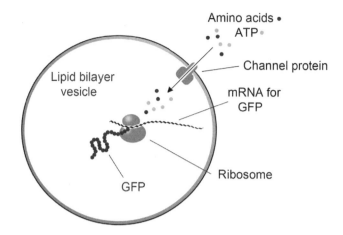

FIGURE 31

In 2004, Vincent Noireaux and Albert Libchaber published a method by which ribosomes could be encapsulated in lipid vesicles. They showed that the ribosomes could synthesize green fluorescent protein (GFP) using messenger RNA that codes for the GFP amino acid sequence. Because lipid membranes are impermeable to substrates, the vesicles also included mRNA for alpha hemolysin, which was synthesized by the ribosomes and formed channels in the membrane. The channels allowed externally added "nutrients" such as amino acids and ATP to enter the vesicles and supply the encapsulated ribosomes with necessary ingredients.

protein synthesis. The researchers then carefully chose two genes to translate, one for green fluorescent protein (GFP), a marker for protein synthesis, and a second gene for a pore-forming protein called alpha hemolysin. If the system worked as planned, the GFP would accumulate in the vesicles as a visual marker for protein synthesis, and the hemolysin would allow externally added "nutrients" in the form of amino acids and ATP, the universal energy source required for protein synthesis, to cross the membrane barrier and supply the translation process with energy and monomers (Figure 31).

The system worked as expected. The vesicles began to glow with the classic green fluorescence of GFP, and the hemolysin allowed synthesis to continue for as long as four days. Tetsuya Yomo and his research group at Osaka University have gone a step further with a similar encapsulated translation system in which the GFP gene is present in a strand of DNA. They refer to their system as a genetic cascade because the GFP gene is transcribed into messenger RNA, which then directs synthesis of the protein.

CONNECTIONS

Can we now synthesize life? This question brings us to the limits of what we can do with our current techniques, and the answer is: not yet. Everything in the system grows and reproduces *except* the catalytic macromolecules themselves. All of the systems to date depend on polymerase enzymes or ribosomes, and even though every other part of the system can grow and reproduce, the catalysts get left behind.

The final challenge is to encapsulate a system of macromolecules that can make more of themselves by catalysis and template-directed synthesis. A little progress has been made to this end. David Bartel and his coworkers at the Whitehead Institute, using a technique developed for selection and molecular evolution, produced a ribozyme that can grow by polymerization, in which the ribozyme copies a sequence of bases in its own structure. So far, the polymerization has only copied a string of 14 nucleotides, but this was a good start, and Peter Unrau at Simon Fraser University recently reported a similar system that could copy sequences up to 20 nucleotides. If a ribozyme system can be found that catalyzes its own complete synthesis using genetic information encoded in its structure, it could rightly be claimed to have the essential property that is lacking so far in artificial cell models: reproduction of the catalyst itself. Given such a ribozyme, it is not difficult to imagine its incorporation into a system of lipid vesicles that would have the basic properties of the living state.

EPILOGUE

Professor Deamer: Is it really the case that taxpayers' money is being wasted on your "soap bubble" theory for the origins of life? This is an insult to the intelligence of the American public as well as an egregious misuse of research money. If you want to rant on and speculate about the importance to biology of self-assembling decanoic acids and the like please do so at your own cost. I will be writing to NASA asking them to terminate assistance for your crass pseudo-science.

FROM AN EMAIL RECEIVED BY THE AUTHOR IN 2008

What is it about research on the origins of life that stirs such strong sentiments in my fellow citizens? I wrote back to this gentleman, saying that I had calculated the cost of my research to U.S. taxpayers, which totaled one cent each. He immediately responded, saying that was one cent too much. I didn't offer to refund his share of the cost, but his response made me realize that in some people's minds, scientific attempts to discover how life began or to create artificial life are a total waste of my time and their tax money, perhaps even blasphemous.

I want to respond to such concerns in an epilogue to *First Life*. This is not intended as a debate with fundamentalists who insist on rejecting what science reveals about the origins of life in favor of what they regard as the true account in the book of Genesis. Their belief system cannot be swayed by scientific evidence, and I can only feel sorry that they are missing the truly wonderful history of life on Earth, which was discovered not by revelation, but by the endless curiosity of scientists who have been asking and answering questions since Galileo's time. Instead, I am writing for thoughtful people who may be puzzled by some topics in this book or who find it incredible that living systems could appear spontaneously on early Earth and evolve into today's diverse biosphere. I am also writing for my fellow teachers who are trying to pass along these ideas to their students without offending religious sensibilities.

Epilogues should be brief, so I will touch on just a few questions that come up time and again in my conversations with students in classrooms and members of audiences when I give talks. I will also discuss the idea of intelligent design, because in the United States this creates major problems when conservative school board members, parents, teachers, and even legislators try to inject their religious beliefs into the classrooms of public schools.

SCIENCE AND RELIGION SEEM TO BE COMPLETELY DIFFERENT WAYS TO UNDERSTAND THE MEANING OF HUMAN EXISTENCE. IS THERE ANY COMMON GROUND?

I agree that there is a fundamental difference in the way that religion and science try to understand our world. For the religious, belief in a supreme intelligence gives a deeper meaning and purpose to their lives. Religious faith also helps many people cope with the insecurity that can arise from the uncertain world we live in and the knowledge that life ends in certain death. Evolutionary biologists have pointed out that this behavioral trait— belief that an unseen, omnipotent entity guides our lives—could have been a positive factor that helped small tribes of early humans survive in a dangerous world. A tribe that cooperates within a shared belief system will be better able to fend off predators than a scattered group of relatively independent individuals. If so, this trait remains embedded in the human nervous system to varying degrees. Some people feel the need to have faith very strongly, others not at all. The different ways that people think about the world can produce a disturbing clash that gives rise to much of the acrimony between science and religion. Imagine that you are asked the following question: Is there a supreme intelligence that made the world and guides our lives? The religious answer is Of course! I am certain of it. The scientific response is more skeptical: Show me the evidence. This scientific response immediately raises the hackles of those who rely on a foundation of faith and certainty to give purpose to their lives. In contrast, the scientific mind is comfortable knowing that nothing is certain and that the best way to understand the world we live in is to ask questions that test what we think we know.

Science and religion can find common ground with the understanding that they are different ways to understand human existence. Both are intrinsically powerful and affect our lives, but neither should lay claim to be the complete and absolute truth. After the fundamental difference is acknowledged, constructive communication can begin.

IT SEEMS IMPOSSIBLE THAT SOMETHING AS COMPLEX AS A LIVING CELL COULD HAVE JUST COME TOGETHER BY CHANCE. EVEN A SIMPLE PROTEIN COULD NOT BE PRODUCED BY A CHANCE PROCESS BECAUSE THE PROBABILITY AGAINST THAT HAPPENING IS ASTRONOMICAL. DOESN'T THAT PROVE THAT THERE WAS A CREATOR WHO CAUSED IT TO HAPPEN?

Early Earth was sterile, with a mixture of water, atmospheric gases, minerals, and non-living organic compounds stirred by several sources of energy. Origins of life researchers propose that there was a process by which these simple compounds spontaneously assembled into the first forms of life, presumably composed of polymers resembling proteins and nucleic acids. Can that possibly be true? This brings us to a calculation that has been used to argue in favor of divine intervention in the origin of life, which is to calculate the probability that a single protein molecule with a specific amino acid sequence could arise by chance. The calculation requires only the most elementary math, so I will repeat it here in order to introduce a second calculation related to self-assembly processes.

Let's consider ribonuclease (RNAse), a relatively small enzyme that hydrolyzes RNA to its component nucleotide monomers. This protein contains 149 amino acids of 20 different kinds, arranged in a specific sequence that allows it to fold into a structure having enzymatic activity. The folded structure of the ribonuclease is determined by its amino acid sequence, and a common way to represent such a sequence is to use a code in which one letter of the alphabet designates each amino acid. The first 10 amino acids in ribonuclease are methionine, glycine, leucine, glutamic acid, lysine, serine, leucine, isoleucine, leucine, and leucine, abbreviated MGLEKSLILL. The entire sequence is shown below:

MGLEKSLILLPLLVLVLAWVQPSLGKETPAMKFERQHMDSAGSSSSSPTY
CNQMMKRREMTKGSCKRVNTFVHEPLADVQAVCSQKNVTCKNGKKNCYKS
RSALTITDCRLKGNSKYPDCDYQTSHQQKHIIVACEGSPYVPVHFDASV

To do the calculation, we start with the probability that a specific amino acid will happen to occupy each position in the sequence by chance. There are 20 possible amino acids, so the probability is 1/20, or 0.05. To calculate the probability of assembling the entire protein by chance, we multiply 0.05 by itself 149 times: $0.05 \times 0.05 \times 0.05. \ldots$ The resulting number is so small that even if we tried one new possible structure every second, the correct structure would not appear by chance in the age of the universe, now taken to be about 13 billion years (4×10^{17} seconds). Therefore, the reasoning goes, divine intervention was required to produce RNAse, as well as the millions of other proteins that are coded by the genomes of all life on Earth.

Now let's compare this to a second calculation in which we consider the possibility that lipid molecules can assemble into a membranous compartment. Imagine that we have prepared a soap solution at pH 8 that is relatively dilute. If you are a chemist, this might be decanoic acid, which is a fatty acid containing 10 carbon atoms: $CH_3-CH_2-CH_2-CH_2-CH_2-CH_2-CH_2-CH_2-CH_2-COOH$. (This substance, of course, is what so exercised my correspondent!) A dilute solution of decanoic acid is perfectly clear. If we could view it at sufficiently high magnification we would see some of the soap molecules free in solution, while others would be in aggregates called micelles. Now let's add soap to make the solution more concentrated. At a certain concentration, something almost magical happens. Some of the soap remains in solution, but the rest assembles into beautiful spherical vesicles, each containing about 100 million soap molecules (Figure 32).

What is the probability that just one such vesicle might assemble by chance? Each soap molecule can either be in the volume of the lipid bilayer that forms the membrane or it can stay in solution. If we assume that the average vesicle is 2 micrometers in diameter, the volume of the vesicle membrane is 6×10^{-4} cubic micrometers. (This is the volume of the membrane itself, not the volume contained within the vesicle.) Each vesicle occupies about 1,000 cubic micrometers in the soap solution, so the chance that a given soap molecule will happen to be in the volume of the vesicle membrane is 6×10^{-4} divided by 1,000, or 6×10^{-7}. To calculate the probability of a 2-micrometer vesicle assembling by chance, we multiply this number by itself 100 million times, which is the number of molecules in the membrane. The resulting number is again so small that I won't even try to write it. Just accept that a single vesicle could not possibly assemble by chance from soap molecules in solution.

And yet, not just a single vesicle, but vast numbers of vesicles appear in the soap solution. This is the power of self-assembly, which readily produces structures that would seem to be

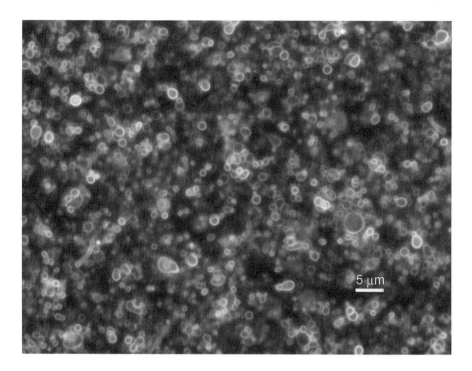

FIGURE 32
When decanoic acid reaches a certain concentration in solution, the molecules self-assemble into the microscopic vesicles shown in this micrograph. The vesicles are in the same size range as typical bacterial cells, and were stained with a fluorescent dye to make them visible.

mathematically impossible. No divine intervention is required, just the presence of soap molecules in solution at a certain concentration. Nothing holds the molecules together in the vesicle except a weak force called the hydrophobic effect. The same force is responsible for stabilizing the soap bubbles we play with as children and the lipid assemblies that form the membranes around every cell in the human body.

In the case of a protein, we are calculating the probability that 20 different amino acids will produce a specific sequence 149 amino acids in length, and with vesicles we are calculating the probability that 100 million soap molecules will happen to occupy a specific volume. *The take-home message is that such calculations can be deceiving, in that they conceal erroneous assumptions.* For lipid vesicles, the error is to assume that the molecules assemble purely by chance, while, in fact, there are physical forces that stabilize the lipid bilayer membrane. In the case of the protein, the error is to assume that the amino acids also assemble by chance, neglecting the fact that any given protein is the product of a long evolutionary process over several billion years in which the sequences emerge from random mixtures not by chance, but by natural selection. The fact that this can happen has been repeatedly demonstrated in the laboratory, one example being the ribozyme experiment described in Chapter 13.

DO SCIENTISTS THINK THAT IT IS WRONG TO BELIEVE IN GOD? IF GOD DOES NOT EXIST, HOW DID THE UNIVERSE BEGIN?

A few outspoken scientists argue that it is wrong to believe in a supreme intelligence, but I don't think this is wrong in the usual sense of the word. In fact, a recent poll has shown that 36% of scientists believe in God, and about half are spiritual in some sense. How does a religious scientist accommodate both the questioning that is characteristic of scientific practice and a belief that there is a supreme intelligence behind it all? I think there is a way for the questioning method of science to meet religious belief without generating the distressing anti-science clash that is newsworthy, yet ultimately destructive. It is clear that matter and energy interact according to a set of physical laws, and strong evidence indicates that the universe had a beginning we call the Big Bang. A religious scientist can comfortably believe that the universe and its laws were put in place by a creator. Science has nothing to say about this statement of faith because it cannot be submitted to experimental tests or observation.

EVOLUTION IS JUST DARWIN'S THEORY AND NOT A FACT. WHY SHOULD WE MAKE KIDS IN SCHOOL LEARN ABOUT SOMETHING THAT IS JUST THEORETICAL? AFTER ALL, NO ONE HAS EVER SEEN SOMETHING EVOLVE OR A NEW SPECIES APPEAR.

The word "theory" has one meaning in ordinary language, but it has a very different meaning in scientific usage. When someone says that an idea is "theoretical," we understand that it is a speculation, little more than a guess or a hunch. But a theory in science is an explanation of a group of related phenomena that can be observed. Furthermore, to advance to the level of an accepted theory, a hypothesis must have been rigorously tested by experiment or observation, and this usually involves a prediction that the theory makes. A classic example is Newton's theory of gravity. Its mathematical expression is that the gravitational force F acting between two objects is proportional to their relative masses (m_1 and m_2) times the gravitational constant G, and is inversely proportional to the square of the distance d between the two objects:

$$F = G(m_1 \times m_2)/d^2$$

The test of Newton's theory was that it predicts the downward force acting on an apple or a bowling ball that we measure as its weight, and it also predicts the force acting between planets and the sun.

Although they are less obvious than gravity, Darwin's theory of evolution also has a set of predictions that have been tested. For instance, even though Darwin had no knowledge of DNA, proteins, or genes, he predicted that there must be a way for genetic information to be transmitted between generations and a mechanism by which variations can occur in the information. When the structure and function of DNA were discovered, it became clear that DNA contained genetic information, and mutations in DNA lead to variations in populations that can then be subjected to natural selection. Darwin's theory is also consistent with the fossil record, which shows that populations of organisms change over time, starting with simple cellular life over three billion years ago, evolving into multicellular life half a billion years ago, and finally resulting in the biosphere today.

Then there is the question of speciation. Most people never see a new species of complex life appear, so it is easy to conclude that it can't happen. But this has a simple explanation: It takes a long time! Only 150 years have passed since Darwin published his most celebrated book, *On the Origin of Species*, but speciation ordinarily requires thousands to millions of years, depending on whether the organism is a fruit fly with a generation time measured in a few weeks, or a bird or mammal with one or more years between generations. For example, the Galápagos Islands are around three to five million years old, and a small population of finches began to live on them shortly after the islands emerged from the sea as small volcanoes, then cooled and became habitable. It is not known how the birds arrived; most likely a flock got caught up in a major storm with winds blowing from South America. What is known is that very slowly, over a period of several million years, that initial flock evolved into 15 separate species, each specialized to eat a certain kind of seed. If Darwin's finches needed several million years to evolve, anyone who expects to see new species appear in a human lifetime is going to be disappointed. But just as we have historical documents to prove that our ancestors once existed centuries ago, the fossil record shows that life on our planet existed more than three billion years ago and undergoes continuous evolutionary change over millions of years that generates new species.

ISN'T IT TRUE THAT SCIENTISTS STILL DISAGREE ABOUT HOW OR EVEN WHETHER EVOLUTION OCCURS? IF WE ARE GOING TO TEACH STUDENTS TO THINK CRITICALLY, THEN WE SHOULD LET THEM LEARN ABOUT ALTERNATIVES TO EVOLUTION, LIKE INTELLIGENT DESIGN. THAT'S ONLY FAIR.

It is true that, like any other scientific theory, the theory of evolution is still a work in progress. Thousands of research papers are published every year by scientists who are trying to understand how it works. Although there are often disagreements about the significance and interpretation of new findings, these are not about whether evolution occurs. There is overwhelming evidence that evolution is by far the best explanation for diversity, speciation, and a fossil record that covers three billion years of Earth's history. The disagreements among scientists about evolution concern the mechanisms underlying evolution and how to modify the theory to take into account new observations. This is a natural part of the scientific process.

In order to sway public opinion, proponents of intelligent design say that because scientists don't seem to agree about evolution, it can't be accepted as a fact. To support their claim, they often produce a handful of scientists who reject evolution. For instance, Michael Behe is a biochemist at Lehigh University who proposes that certain structures in living organisms must have been designed. One example is the way that some bacteria move. Microscopic motors called flagella are embedded in their membranes, and protons moving through the motors cause the flagella to rotate like tiny propellers. Behe calls this irreducible complexity, which basically states that irreducibly complex systems or processes can't function without all of their parts in place. Because natural selection can only favor functional systems and processes, it follows that natural selection can't produce irreducibly complex systems or processes. In other words, the flagellar motor and mechanism are so complex, and the working parts so perfect, that they could not possibly have arisen by evolutionary processes.

Because he is a scientist with a PhD, Behe's ideas have been promoted as a scientific alternative to evolutionary theory that should be taught to students in science classes. And in general, proponents of intelligent design have often argued that students should be taught about intelligent design in public schools so that they can choose for themselves. These arguments are often popular, especially in the United States. The American population is the most religious culture among western societies. Poll after poll shows that over half of all Americans accept the Bible as literal truth and reject the fundamental concepts of evolution. It is only natural that parents want their children to believe the same things they do, so they are disturbed when a science teacher teaches evolution as though it were a fact. Because of this, certain citizen groups believe that evolution should be considered to be "only a theory," and that students should be given the opportunity to consider intelligent design as an alternative explanation.

This sounds fair, but it really isn't because intelligent design is an opinion that has no scientific basis. Those who believe in intelligent design say that that bacterial flagella are obviously too complex to have arisen by a process of evolution, that they must have been designed. Instead of adding to our knowledge, this effectively ends the questioning process that is fundamental to science. Even when the proposal is simply to teach that evolution is scientifically controversial, it still isn't fair. There is nothing fair about misleading students to think that evolution is scientifically controversial when, in fact, it is not.

Moreover, teaching intelligent design in the public schools is itself a violation of American ideals of justice. The First Amendment to the United States Constitution has a provision called the Establishment Clause, which basically says that religious belief must be kept separate from government, that there must be a separation of church and state. For me, this is one of the most important clauses in the Constitution, but some citizens, particularly creationists and intelligent design proponents, disagree with the Establishment Clause and believe that their particular religious faith, thinly disguised as science, should be included in public education.

In 2008, proponents of intelligent design had a chance to make their case in a trial in Dover, Pennsylvania. Creationists who had managed to get elected to the Dover school board told teachers that they must call attention to intelligent design as an alternative to evolution. They also suggested that a creationist-oriented book called *Of Pandas and People: The Central Problem of Biological Origins* should be recommended to students. Eleven parents challenged this policy and were able to bring their case to court. Michael Behe and others were there as expert witnesses to argue in favor of intelligent design, and biologist Kenneth Miller and others argued in favor of the plaintiffs, the parents who had brought the suit. The presiding judge was John E. Jones III, who had been appointed by President George W. Bush.

The Dover trial was very dramatic, making national and international headlines. The drama was reminiscent of the 1925 trial in which John Scopes, a high school biology teacher in Tennessee, was accused of violating state law when he taught his students that evolution was a fact. The Dover trial took 40 days of testimony from parents, board members, and the expert witnesses, after which Judge Jones ruled in favor of the parents. It is worth quoting a few summary lines from his ruling (in which he refers to intelligent design with the abbreviation ID):

> After a searching review of the record and applicable case law, we find that while ID arguments may be true, a proposition on which the Court takes no position, ID is not science. We find that ID fails on three different levels, any one of which is sufficient to preclude a determination that ID is science. They are: (1) ID violates the centuries-old ground rules of science by invoking and

permitting supernatural causation; (2) the argument of irreducible complexity central to ID, employs the same flawed and illogical contrived dualism that doomed creation science in the 1980's; and (3) ID's negative attacks on evolution have been refuted by the scientific community.

[T]he one textbook [*Pandas*] to which the Dover ID Policy directs students contains outdated concepts and flawed science, as recognized by even the defense experts in this case.

ID's backers have sought to avoid the scientific scrutiny which we have now determined that it cannot withstand by advocating that the controversy, but not ID itself, should be taught in science class. This tactic is at best disingenuous, and at worst a canard. The goal of the IDM is not to encourage critical thought, but to foment a revolution which would supplant evolutionary theory with ID.

Accordingly, we find that the secular purposes claimed by the Board amount to a pretext for the Board's real purpose, which was to promote religion in the public school classroom, in violation of the Establishment Clause.

SHOULD SCIENTISTS BE ALLOWED TO TRY TO SYNTHESIZE ARTIFICIAL LIFE? IF THEY SUCCEED, ISN'T THERE A DANGER THAT THEY MIGHT ACCIDENTALLY MAKE SOMETHING LIKE A VIRUS THAT COULD ESCAPE AND PRODUCE A PANDEMIC?

I think that we will soon be able to use our knowledge of physics, chemistry, and biology to assemble an artificial living system, as described in Chapter 15. The reason is that a deep understanding of life's mechanisms has emerged over the past 50 years. For instance, we now know most of the sequence of three billion base pairs in the human genome, an achievement that was undreamed of when Watson and Crick established the double helix structure of DNA in the early 1950s. We can manipulate bacterial genomes at will, cause stem cells to develop into specific tissues, and even produce clones of sheep, dogs, cats, and other mammals. Given the momentum of scientific progress, it seems entirely plausible that in the next 50 years, we will be able to reassemble bacterial cells from a parts list and perhaps even produce a new form of life—a second origin of life, but this time under laboratory conditions.

Researchers at the J. Craig Venter Institute have estimated that 382 genes is the minimal number required for a simple bacterium to grow and reproduce, and while this book was being written, Venter's team reported that they had successfully transferred a complete synthetic genome from one bacterial species to another. As often occurs with powerful new knowledge, their achievement generated considerable controversy: Is it ethical to attempt to synthesize a living organism? Should there be laws against tinkering with life this way?

The immediate concern, of course, is the balance between valuable new knowledge and public safety. On the plus side, we might be able to design artificial forms of life that can manufacture inexpensive foods, biofuels, and therapeutic agents. For example, Genentech already uses genetically modified bacteria to make human insulin and other proteins, and most major pharmaceutical industries brew up huge vats of microorganisms to produce antibiotics. But these are only slightly modified versions of existing forms of life, and it is a costly and complex process to isolate a small amount of desired product from a mass of bacterial protein. The goal of synthetic biology is to design a simplified version of life that produces the desired product with high efficiency and uses an abundant energy source such as sunlight, or cellulose derived from agricultural waste.

So far, there are virtually no restrictions on attempts to synthesize artificial life nor should there be, in my judgment. But bioethical principles require us to look further, to weigh risks and benefits. What are the risks of learning how to make artificial life? Given the example of HIV and the AIDS epidemic, it's not difficult to look ahead to potential risks. The public, the taxpayers who support most research, will wonder whether a highly infectious artificial organism might escape the laboratory. As often happens, writers of science fiction have examined such possibilities. Michael Crichton describes one such scenario in his novel *Prey*, in which nanoscopic robots designed to gather military intelligence escape from the laboratory and evolve into predatory swarms.

Even worse than an accidental release, what about the possibility that a terrorist group will design a synthetic bacterium that is passed between human beings, multiplies in the gut, and produces botulin toxin? Or what if a virus like HIV is produced that can be transmitted like influenza, rather than as a sexually transmitted disease? Just imagine the public outrage if the AIDS virus was discovered to be someone's invention, rather than an example of viral evolution in the wild which also happened to be infectious in humans. Shouldn't there be a law against tinkering like this?

Global society is too complex ever to exert absolute control over attempts to produce artificial life. An example is stem cell research. For political reasons, the George W. Bush administration banned the use of federal funding to develop new embryonic stem cell lines. South Korea simply took this as an opportunity to jump ahead in the race of discovery. What we should do is to look ahead to the potential dangers and try to provide thoughtful guidance to researchers so that they will consider the dangers and be aware of public concerns. Most research in this area is supported by grants from federal agencies such as NIH, NSF, and NASA, and all proposals undergo peer review. We must make this process as transparent as possible and have open discussions of risks and benefits so that the peer review panels, who decide which proposals should be funded, can make informed decisions about the value of the research and potential dangers. The search for a method to assemble a synthetic version of life is still in an initial curiosity-driven phase, and there is no reason to do anything but cheer on the few scientists who are bravely exploring this new territory.

SOURCES AND NOTES

In the scientific literature, most references are to papers published in journals, rather than Internet websites. However, journals are generally available only in university libraries, so most of the sources listed below refer to more easily accessible websites. A few particularly relevant papers and books are also cited.

CHAPTER 1

Astrobiology has attracted international attention among scientists. You can learn more about this emerging field by visiting the following websites:

http://astrobiology.nasa.gov/
http://www.astrochem.org/
http://www.esa.int/

Kevin Plaxco, at the University of California, Santa Barbara, with science writer Michael Gross recently published a very readable book called *Astrobiology: A Brief Introduction* (2006). Another remarkable book is *Rare Earth: Why Complex Life Is Uncommon in the Universe* by Peter Ward and Don Brownlee at the University of Washington (2000). They convincingly argue that the origin of life on Earth and its evolution toward intelligence might be much rarer than we think.

It is well worth visiting the collection of Hubble Space Telescope images at the *Hubble Site* (http://hubblesite.org/gallery) There you will find magnificent views of exploding stars and planetary nebulas, supernovas, interstellar dust clouds giving birth to stars, and jewel-like spiral galaxies. Another great website is *Astronomy Picture of the Day*, sponsored by NASA (http://apod.nasa.gov/apod). Every day a new image is posted from the Hubble and other telescopes either in orbit or in places far from city lights like Mauna Kea in Hawaii or the Atacama Desert in Chile.

For artists' versions of what was discussed in this chapter, see the websites of William Hartmann (http://www.psi.edu/~hartmann/painting.html) and Don Dixon (http://cosmographica.com/gallery).

Readers can learn more about stellar nucleosynthesis at a website hosted by Glen Elert called *The Physics Hypertextbook* (http://physics.info/nucleosynthesis), which clearly describes how biogenic elements are synthesized in stars.

For a scientific account of meteorites, Cambridge University Press published *Meteorites: A Petrologic, Chemical and Isotopic Synthesis* by Robert Hutchison (2007). On the *Scientific American* website, you find a recent podcast called "Murchison Meteorite's Chemical Bonanza" (http://www.scientificamerican.com).

Sponsored by the American Institute of Physics, a website called *Ideas of Cosmology* (http://www.aip.org/history/cosmology/ideas/expanding.htm) gives an excellent history of how the competing ideas of George Gamow and Fred Hoyle were finally resolved in favor of the Big Bang over a steady state universe. To learn more about the Big Bang theory, see NASA's website for the Wilkinson Microwave Anisotropy Probe (http://map.gsfc.nasa.gov/universe).

A website sponsored by the Harvard University Department of Astronomy (http://www.cfa.harvard.edu/COMPLETE/learn/star_and_planet_formation.html) describes how solar systems are produced in interstellar molecular clouds. Several links to animations of this process are also given.

Brig Klyce has an interesting website called *Cosmic Ancestry* that presents arguments in favor of panspermia to explain the beginning of life on Earth (http://www.panspermia.org/index.htm). The ideas of Fred Hoyle and Chandra Wickramasinghe are discussed in some detail there.

CHAPTER 2

For further information about volcanoes, see *Global Volcanism Program* (http://www.volcano.si.edu) sponsored by the Smithsonian Institution. There you can search for volcanoes all over the world, including Kamchatka. For instance, if you search for Mutnovski, you will find information about and photographs of the magnificent volcano that I described at the beginning of this chapter.

"Self-Assembly Processes in the Prebiotic Environment," the scientific report we wrote about our research in Kamchatka was published in 2006 in *Philosophical Transactions of the Royal Society B.*

A nice introduction to hydrothermal vents is found on the website *Dive and Discover*, sponsored by Woods Hole Oceanographic Institution (http://www.divediscover.whoi.edu/vents/index.html).

CHAPTER 3

In 2005, John Valley summarized his studies of zircons in an article for *Scientific American* entitled "A Cool Early Earth?" The article clearly describes how the elemental composition of zircons reveals their age and the temperature at which they form. A pdf file is available at the

University of Wisconsin–Madison Department of Geoscience website (http://www.geology
.wisc.edu/zircon/Valley2005SciAm.pdf).

The website *Meteor Crater* (http://www.meteorcrater.com) is sponsored by the company
that runs tours into the crater. It is very informative and includes an impressive video presen-
tation of what the impact looked like when a huge iron meteorite struck the Arizona desert
50,000 years ago.

The book *Planets and Life* edited by W. T Sullivan and J. A. Baross (2007) provides a broad
introduction to the new field of astrobiology. Chapter 12, written by Roger Buick, discusses
bacterial microfossils and offers a thoughtful overview of how they are formed and how much
confidence we can have that they are real.

CHAPTER 4

Chemistry and biochemistry are particularly well-served by Internet sites. Simply typing a
chemical name into your browser will usually take you directly to a source, often *Wikipedia*,
that contains basic information.

As a public service, the National Center for Biotechnology Information has made over
200 scientific and medical books freely available at the center's website (http://www.ncbi
.nlm.nih.gov/books). One of these books is *Molecular Biology of the Cell, Fourth Edition*, written by a team of scientists led by Bruce Alberts (2002). In that book,interested readers will
find detailed information about the biomolecules discussed in Chapter 4.

CHAPTER 5

The Exploring Space section of the Public Broadcasting Service (PBS) website (http://www
.pbs.org/exploringspace/meteorites/murchison/index.html) has a nice introduction to meteorites and the origin of life, including a discussion of how the chirality of amino acids could
have played a role in initiating homochirality.

Anne Marie Helmenstine hosts a searchable website *About.com: Chemistry* (http://
chemistry.about.com) that presents fundamental chemical concepts at an introductory level.
A search for "chirality" will provide further details about the chiral nature of amino acids and
other biomolecules.

CHAPTER 6

The Department of Energy has an excellent introduction to concepts of energy on its website
(http://www.eia.doe.gov/kids). Although it's called *EnergyKids*, it is an award-winning website
that anyone can learn from.

Franklin Harold spent his scientific career investigating how simple forms of life use
energy. Frank wrote a nontechnical book called *The Vital Force: A Study of Bioenergetics* (1986).
Although the book was written more than 20 years ago, it remains one of the most readable
and lucid introductions to the role of energy in life processes.

With coauthor Art Weber, I wrote a chapter of the book *Origins of Life* (2010) that discusses the kinds of energy available to drive chemical reactions leading to the origin of life.

The chapter is available in the 2010 archives of the Cold Spring Harbor Press website (http://cshperspectives.cshlp.org).

Quite a few video animations of electron transport and ATP synthesis have been posted on the *YouTube* website (www.youtube.com) and can be found by using those words as search terms. One that I enjoy watching and that I often use in class is called "grt video of atp synthase."

CHAPTER 7

An excellent website describing fundamental aspects of self-assembly processes is hosted by the Materials Research Science and Engineering Center at the University of Wisconsin–Madison (http://mrsec.wisc.edu/Edetc/nanoquest/self_assembly/index.html). The website includes several videos illustrating self-assembly processes.

CHAPTER 8

Exploring Life's Origins (http://exploringorigins.org/) is a superb website developed by Janet Iwasa in collaboration with Jack Szostak. The illustrations and animations clearly illustrate the main points of this chapter, including the self-assembly of lipid vesicles, encapsulation of nucleic acids, and growth by polymerization. The website for Encapsula NanoSciences (http://www.encapsula.com) has posted two informative videos on *YouTube* (www.youtube.com) that give clear descriptions of lipids, liposomes, and how lipid vesicles can encapsulate solutes.

The First Cell Membranes, a review summarizing my research on encapsulation and the origin of cellular life, was published in *Astrobiology* in 2002. A pdf file can be downloaded from the website hosted by the Astrophysics & Astrochemistry Laboratory at NASA Ames Research Center (http://www.astrochem.org/PDF/Deameretal2003.pdf).

CHAPTER 9

The Institute for Systems Biology in Seattle, Washington, hosts an informative website that introduces basic aspects of this emerging discipline (http://www.systemsbiology.org/Intro_to_ISB_and_Systems_Biology).

In addition to his interest in bioenergetics, Franklin Harold was among the first to recognize that life is best understood as a system of interacting molecules. Harold put his ideas together in his book *The Way of the Cell: Molecules, Organisms, and the Order of Life* (2001).

CHAPTER 10

Wikipedia is an obvious resource for all of the subjects in this book, but I have avoided referring to it except for my discussion of biopolymers in Chapter 10. The fact is that the website's articles on proteins, DNA, and RNA are as good, or better, than you will find in any biochemistry textbook. So if you want to know more about biopolymers, *Wikipedia* is the place to start (www.wikipedia.com).

I can recommend two animations of ribosome function that are both interesting and informative. They also provide a contrast between what was possible in 1970 and what can be done now with computer animations. The first can be found on *YouTube* (www.youtube .com) with the title "Protein Synthesis: An Epic on the Cellular Level." In that video, Nobelist Paul Berg at Stanford University begins by discussing how messenger RNA interacts with ribosomes to synthesize protein. The video then segues to a movie showing hundreds of Stanford undergraduates performing a kind of dance in the quad that illustrates what Paul was talking about. The second video, with the *YouTube* title "Transcription and Translation," is from a recent PBS program called *DNA: The Secret of Life*. This is truly spectacular. The molecules are shown as molecular structures with correct relative dimensions, and the computer animation illustrates transcription and translation occurring at the rates they do in a cell.

In Chapter 10, I described how music can be used as a teaching tool to understand ribosome function. If readers are interested, type "dna music" into an Internet browser to discover dozens of websites exploring translation of base sequences into musical notation. These include the website for the Nucleic Acid Database, the primary scientific database used by scientists to store genetic information (http://ndbserver.rutgers.edu/atlas/music/index.html).

CHAPTER 11

John Kimball wrote and published a very successful college-level biology text in 1965, which is now in its fifth edition. The latest version is available online at *Kimball's Biology Pages* (http:// users.rcn.com/jkimball.ma.ultranet/BiologyPages). This is a searchable website, updated frequently, which has basic information about all of biology, including biochemistry. Although the entries are necessarily brief, they are clearly presented and linked to other pertinent entries. For instance, if the word "enzymes" is typed into the search function, the reactions of catalase and carbonic anhydrase are described, with links to cofactors, inhibition, a model enzyme, and enzyme regulation. The Alberts text referred to earlier as a source for Chapter 4 can be consulted for more detailed information about enzyme catalysis and metabolism.

CHAPTER 12

FreeScienceLectures has a 2-minute video animation posted on *YouTube* (www.youtube.com) with the title "DNA Replication Process." The animation clearly shows how polymerase III copies the leading strand directly and then uses RNA primers, Okazaki fragments, DNA polymerase I, and DNA ligase to copy the lagging strand. Also posted on *YouTube* from the PBS show *DNA: The Secret of Life* is an animation called "DNA Replication" that again illustrates the extraordinary complexity of this reaction.

The website for the Dolan DNA Learning Center at the Cold Spring Harbor Laboratory (http://www.dnalc.org) is a wonderful resource for learning about the many functions of DNA in living cells. One of their videos is posted on *YouTube* with the title "Polymerase Chain Reaction (PCR)." Also on *YouTube* you will find a remarkable music video, sponsored by Bio-Rad Laboratories, with the title "Polymerase Chain Reaction (PCR) Song," in which a chorale sings a hymn dedicated to PCR!

CHAPTER 13

To paraphrase a comment about the weather, "Everyone talks about Darwin, but no one reads his book!" During one of my sabbaticals in England, I was nosing around in a used-book store and happened to find *The Origin of Species (6th edition)* for sale for 12 pounds. The book was published in 1902, and my copy was first purchased by Sydney Applegate on December 16, 1904. (Sydney's autograph and date are on the inside cover.) I confess that I have not read Darwin's book cover to cover, but I have read it as I do most scientific books and feel fortunate to have it in my library. Readers can find the complete works of Darwin posted on the website *Darwin Online* (http:// darwin-online.org.uk). This website is well worth visiting because it is an amazing collection that includes Darwins's notebooks and drawings, as well as his publications.

Understanding Evolution is a website sponsored by the University of California, Berkeley (http://evolution.berkeley.edu/evolibrary/home.php). The language is intended for nonscientist readers, but the information content is excellent. One of the links, From soup to cells—the origin of life, gives a nutshell description of the topics covered in this book.

CHAPTER 14

The ideas and concepts in Chapter 14 are presented in earlier chapters of the book.

CHAPTER 15

The Massachusetts Institute of Technology (MIT) has pioneered the use of Internet videos as an educational medium. One such video is a nice lecture by Drew Endy, whose research specialty is synthetic biology. Readers can watch the lecture at MIT's website (http://mitworld .mit.edu/video/363).

Another very nice program about synthetic life can be viewed at the PBS website (http:// www.pbs.org/wgbh/nova/sciencenow/3214/01.html). Hosted by Robert Krulwich, the program features the research of several scientists, including some described in this book.

There are many other books on the origins of life that I recommend reading. I used quite a few as sources of information while writing this book, so it seems worth listing them here for interested readers. Most are available at Internet book dealers such as Amazon (www.amazon.com).

The Molecular Origins of Life: Assembling Pieces of the Puzzle, edited by André Brack (1998)
Genetic Takeover and the Mineral Origins of Life, by Graham Cairns-Smith (1982)
Seven Clues to the Origin of Life: A Scientific Detective Story, by Graham Cairns-Smith (2000)
First Steps in the Origin of Life in the Universe, edited by Julián Chela-Flores, Tobias Owen and
  Francois Raulin (2001)
The Origin of Life, by Paul Davies (2006)
Genesis on Planet Earth: Search for Life's Beginning, by William Day (1984)
Blueprint for a Cell: The Nature and Origin of Life, by Christian DeDuve (1991)
Origins of Life, by Freeman Dyson (1999)
The Origin of Life: A Warm Little Pond, by Clair Edwin Folsome (1979)
Emergence of Life on Earth: A Historical and Scientific Overview, by Iris Fry (2000)
Genesis: The Scientific Quest for Life's Origins, by Robert Hazen (2005)

Biogenesis: Theories of Life's Origins, by Noam Lahav (1999)

The Origin and Early Evolution of Life: Prebiotic Chemistry, the Pre-RNA World, and Time, by Antonio Lazcano and Stanley Miller (1996)

The Emergence of Life: From Chemical Origins to Synthetic Biology, by Pier Luigi Luisi (2010)

The Origins of Life on the Earth, by Stanley Miller and Leslie Orgel (1974)

Beginnings of Cellular Life: Metabolism Recapitulates Biogenesis, by Harold Morowitz (2004)

Origin of Life, by Aleksandr Oparin (1938)

Between Necessity and Probability: Searching for the Definition and Origin of Life, by Radu Popa (2004)

Planetary Systems and the Origins of Life, edited by Ralph Pudritz, Paul Higgs, and Jonathon Stone (2007)

Chemical Evolution and the Origin of Life, by Horst Rauchfuss (2008)

Life's Origin: The Beginnings of Biological Evolution, by J. William Schopf (2002)

Origins: A Skeptic's Guide to the Creation of Life on Earth, by Robert Shapiro (1986)

INDEX

adenine, 69

adenosine triphosphate (ATP), 93, 94, 199, 232

age of the universe, 17

Allamandola, Louis, 61, 74

alpha helix, 172

Altman, Sidney, 191

amino acid, 6, 58, 67, 162

amphiphiles, 122, 133

Anders, Edward, 70

anhydride bond, 142

apatite, 93

Apel, Charles, 120

Archaean eon, 38

Armstrong, Peter, 134

Arrhenius, Svante, 15

Astbury, William, 171, 174

asteroid, 8

astrobiology, 2

atmosphere, 28, 66

atomic number, 40

atomic weight, 40

ATP synthase, 108

Awramik, Stanley, 46, 47

Bada, Jeffrey, 34, 88

Baltscheffsky, Herrick, 232

Bangham, Alec, 111

Barghoorn, Elsa, 46

Baross, John, 27

Barringer crater, 44

Bartel, David, 193, 214–215

base pairs, 206

Benner, Steve, 69

Berg, Howard, 109

Bernal, Desmond, 179

Bernstein, Max, 74

beta sheet, 172

Big Bang, 13

Composition: Michael Bass Associates
Text: 9.5/14 Scala
Display: Scala Sans
Printer and Binder: Thomson-Shore